Poser

A sketchbook of ideas
for artists and models

All images created by Timothy B. Anderson

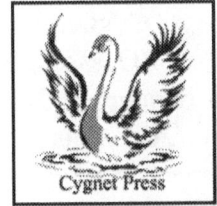

Cygnet Press

also by Timothy B. Anderson:
Frame of Mind
        a collection of poetry

ISBN: 978-0-6151-4657-7

Cygnet Press
P.O. Box 3941
Albuquerque, NM 87190
info@cygnetpress.com

Foreword

This book arrives at the right moment: most art schools do not focus anymore on sketching the Nude, as well as doing black-and-white photographs, while these two mediums are the basics of art. Look at the sketch books of Picasso, Degas, or Matisse! One day, Jean Cocteau, visiting the Picasso Museum in Antibes, had the great surprise to see Matisse sitting just across from the large "NUDE" by Picasso, with a sketch book and a pencil in his hands. At this point a couple of visitors who did not recognize the maestro asked him, "What do you do, sir?" and the answer was, "As you do, I try to understand."

With this corpus we will much better understand the mystery of the human figure—female or male—and do our best to follow the path of Matisse. It is so complicated to pose undressed in front of artists, while they are dressed. And when it is a photographer, the camera and long lens can look like weapons! The model has to be respected and celebrated. We can see how Tim has been looking carefully with great respect to the one who gives us the great privilege to admire the body in its pure perfection. A model is equal to the artist. She expresses herself as well as we do with our medium, and this is such a miraculous exchange between us who will create this "children of soul," together.

Looking to this group of sketches, I am ready to go to work, and I encourage all readers to do the same, with such good assistance, with such a guide. Tim is someone who knows his material. After all, his experience is in the field of art.

Lucien Clergue, Arles, France
March, 2007

Introduction

It is a well-known fact that drawing the human figure has been a creative pursuit for thousands of years. Renowned painters refined this endeavor over the centuries. Since capturing an object on a metal plate came to bear in the mid-nineteenth century, and the ensuing discoveries that led to modern photography became commonplace, the nude figure has persevered as a challenge to all artists, photographers included. Perhaps that is because the human body has more hidden corridors than almost any other object: the corners of an elbow, the abstract lines of the collarbone, the shadows hiding just beneath a muscle, or the soft curves of a breast. These all compete for the eye of the artist. All human bodies are different and no two artists look at them with the same pair of eyes.

The idea for this book came out of my personal work with nude and/or figure models. In the last few years I have made an attempt to strengthen my images, to create pictures that were at once captivating and compelling. To that end, I have worked with a wide variety of models who each brought a separate set of perceptions and ideas about what they wanted to achieve.

In my experience, one of the best things to do when you have contacted a potential model is to suggest meeting for a cup of coffee or tea. It is also helpful if you bring your portfolio, a book, or a magazine with images you want to use as a stepping stone for the shoot. You can each go over what you expect to gain from the session. You can even see if you like each other. On more than one occasion, when I haven't "done what I preached," it has come back to haunt me. This usually happens within two frames of our scheduled session. A pre-shoot meeting also allows you to share portfolios, wardrobe possibilities, makeup, and accessory ideas.

Many times, especially with inexperienced models (and artists), there is a period of time at the beginning of the shoot when the photographer and the model can spend a good deal of time trying to figure out how to do the shoot. Then, frequently, by the time the shoot is completed not too many good pictures are made. This book is an effort to "soften" but not eliminate the initial effort of trying to "get to know each other."

In my sessions, when the model is ready to begin, I ask her if she is comfortable, and if can get her anything: coffee, water, soda, etc. I

also ask her if she likes a particular kind of music. These questions
tend to relax the model and enable her to get into the mood of the
shoot. When we are both ready, I have her disrobe and begin to turn
around. I do this so I can get an idea of how I want to integrate my
plan for the shoot into what her body "shows" me. Since we will have
(ideally) already met and shared portfolios, we should both be starting
on the same page.

What this book is not, is a technical treatise on how one should
pose a model, or what kind of equipment to use. It is, rather, just as it
offers in the title: a book of creative ideas for artist and model alike. I
have chosen to present *Poser* in a manner that illustrates poses without
a lot of clutter, seeking to focus on the pose, itself. Using a sketch
approach allows an image maker to see and study the pose and
develop ideas from that point. I have also included a few of my
photographs at the end of each section in order to expand upon some
of the details of the images in the book.

Also in this book you will find a variety of commentary by
professional figure/nude photographers and models. Rather than use
this book as a method by which I could present my personal
philosophy of nude photography, I decided to discover unique
insights from those who do it every day–the shooters and the models–
and share those guidelines with you. There is biographical and
contact information for each contributor at the back of the book. In
addition to the contributors, whose work I greatly admire, I have
included a list of books I have personally used on many occasions for
direction and inspiration.

Use the images in *Poser* as a launching pad for a new direction
or fresh ideas in your desire to become a better photographer or artist
of the human figure. If you only get one sound idea, then I will
consider the book to be a success.

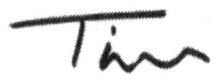

Table of Contents

Light & Shadow

This is what it is all about in the world of creative imagery. Recognizing the play of light and shadow is integral to our development as artists. How a shadow falls across a body is, at times, magic in and of itself. There have been quite a few instances when I have put my eye to the viewfinder only to be held speechless in regard to what I saw. It could have been the way sunlight drew an opaque line across the body of my model, or maybe a shadow was used to obliterate an unwanted part of the image. In any case, whether you use studio strobes, a flash attached to the body of your camera, or natural light, pay attention to what the light is (or is not) telling you.

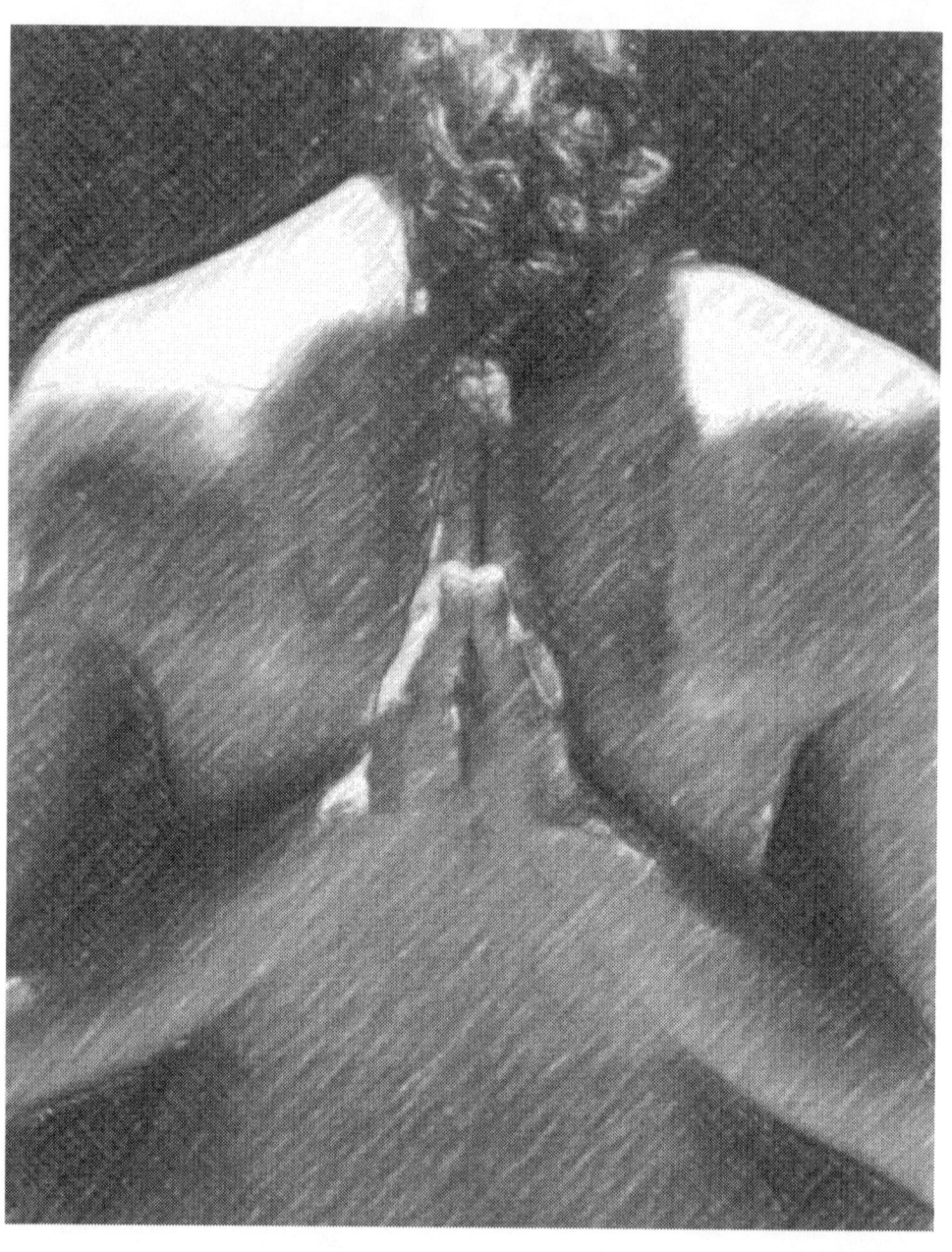

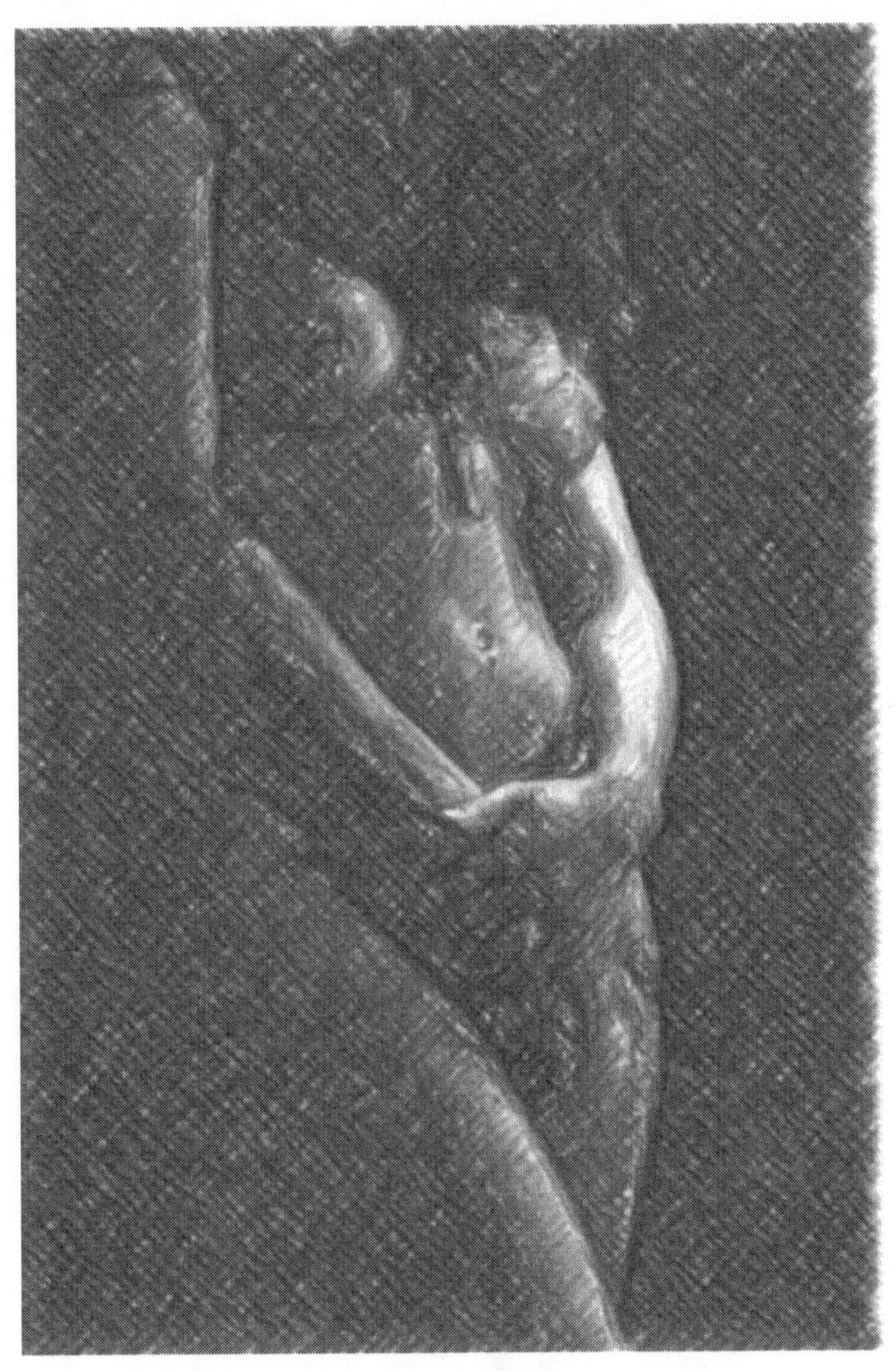

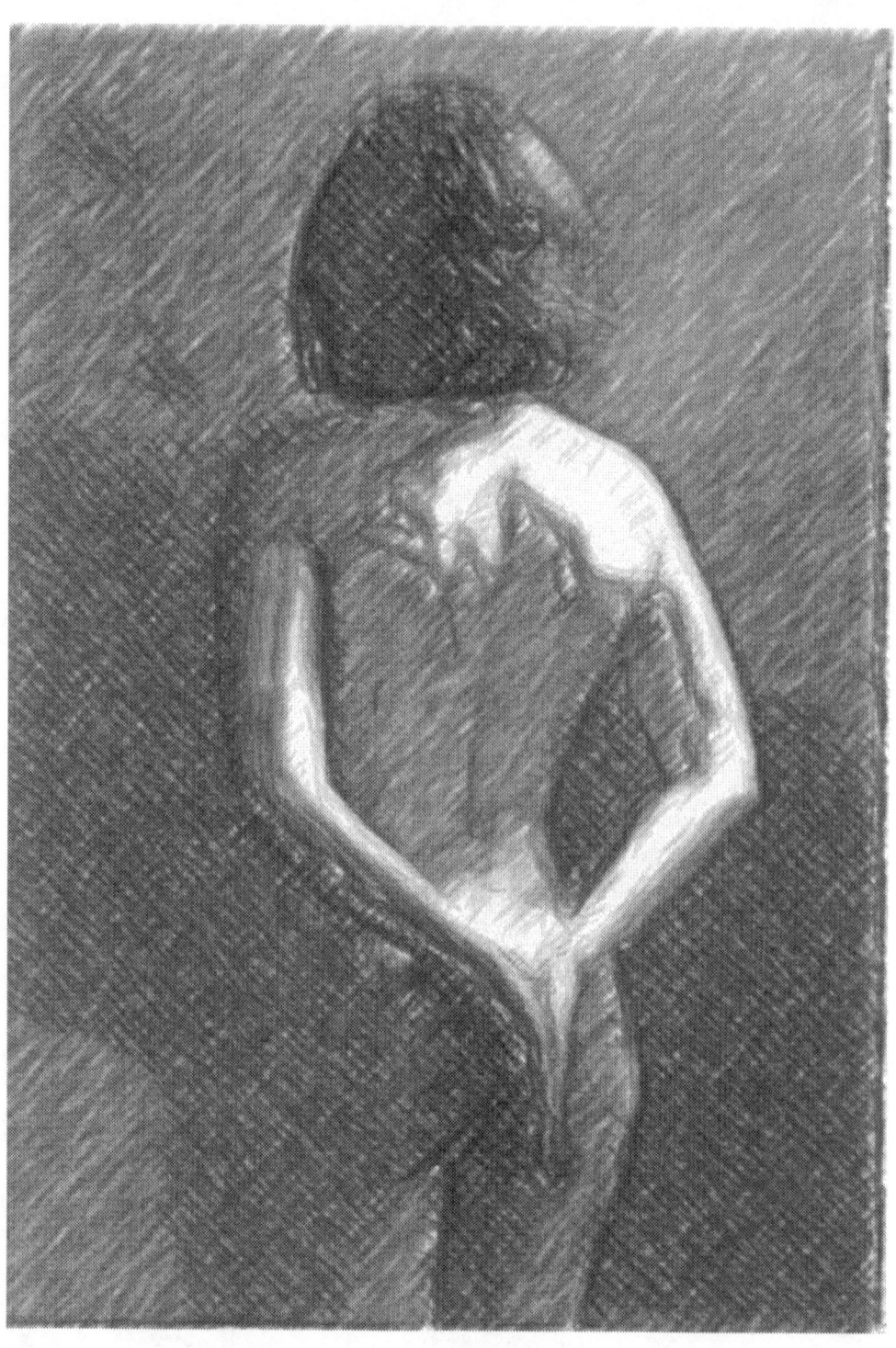

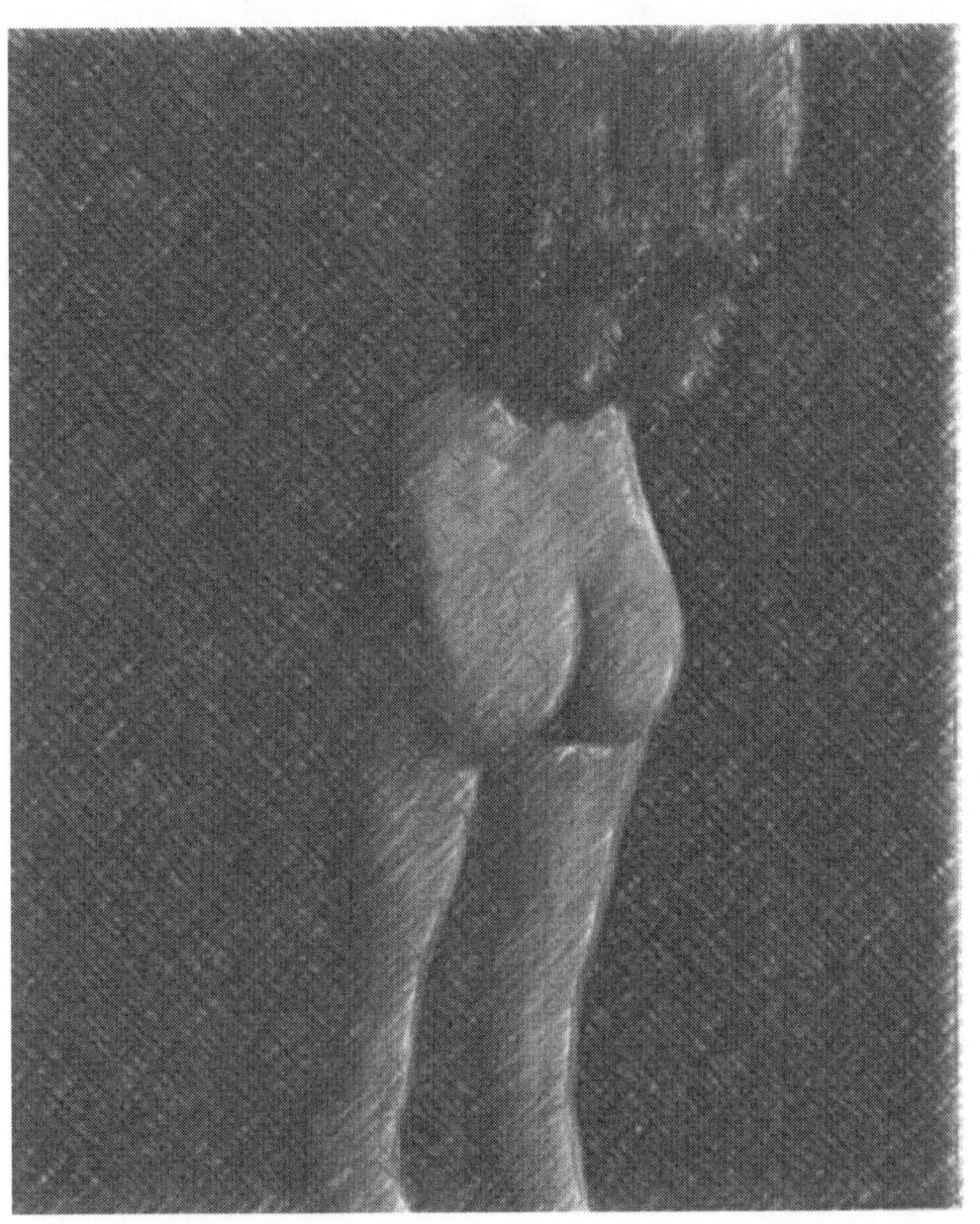

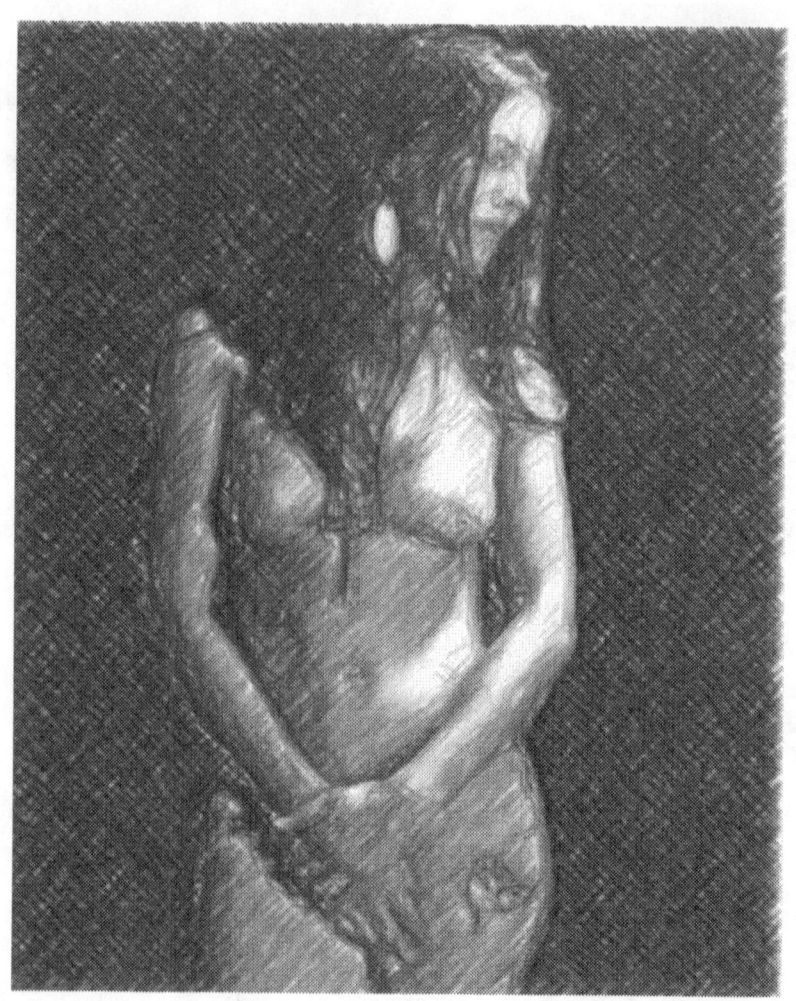

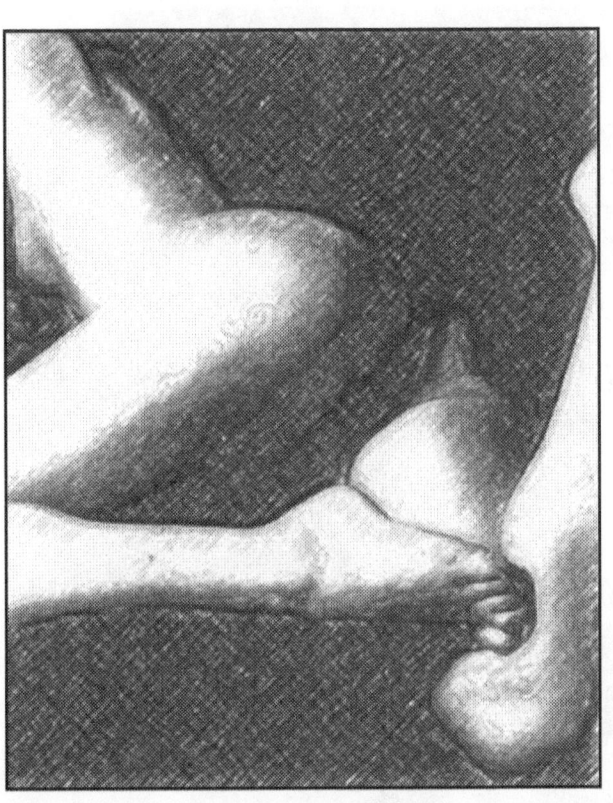

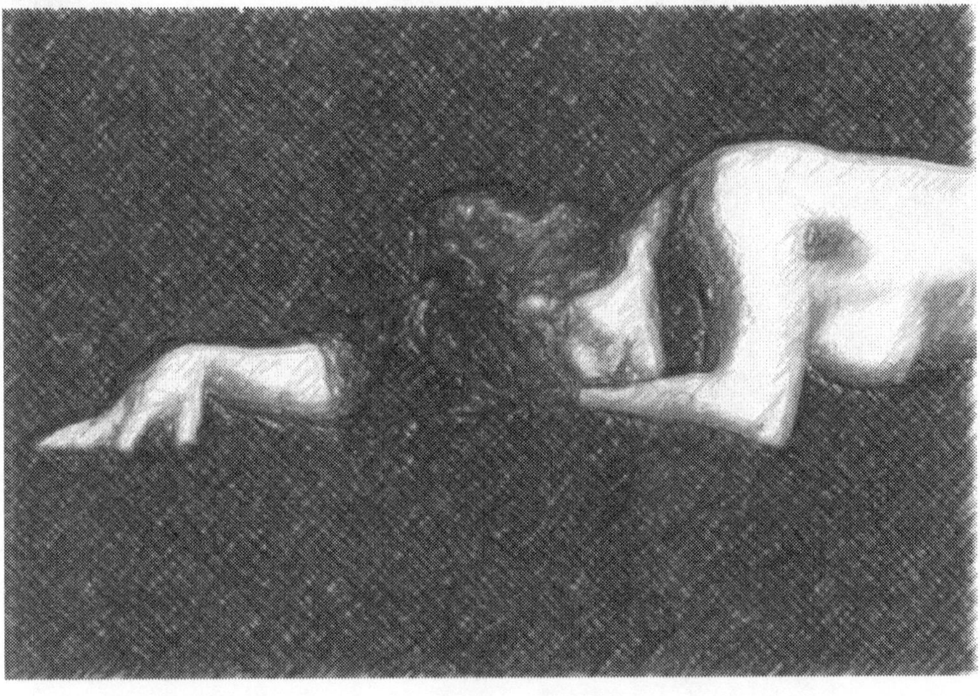

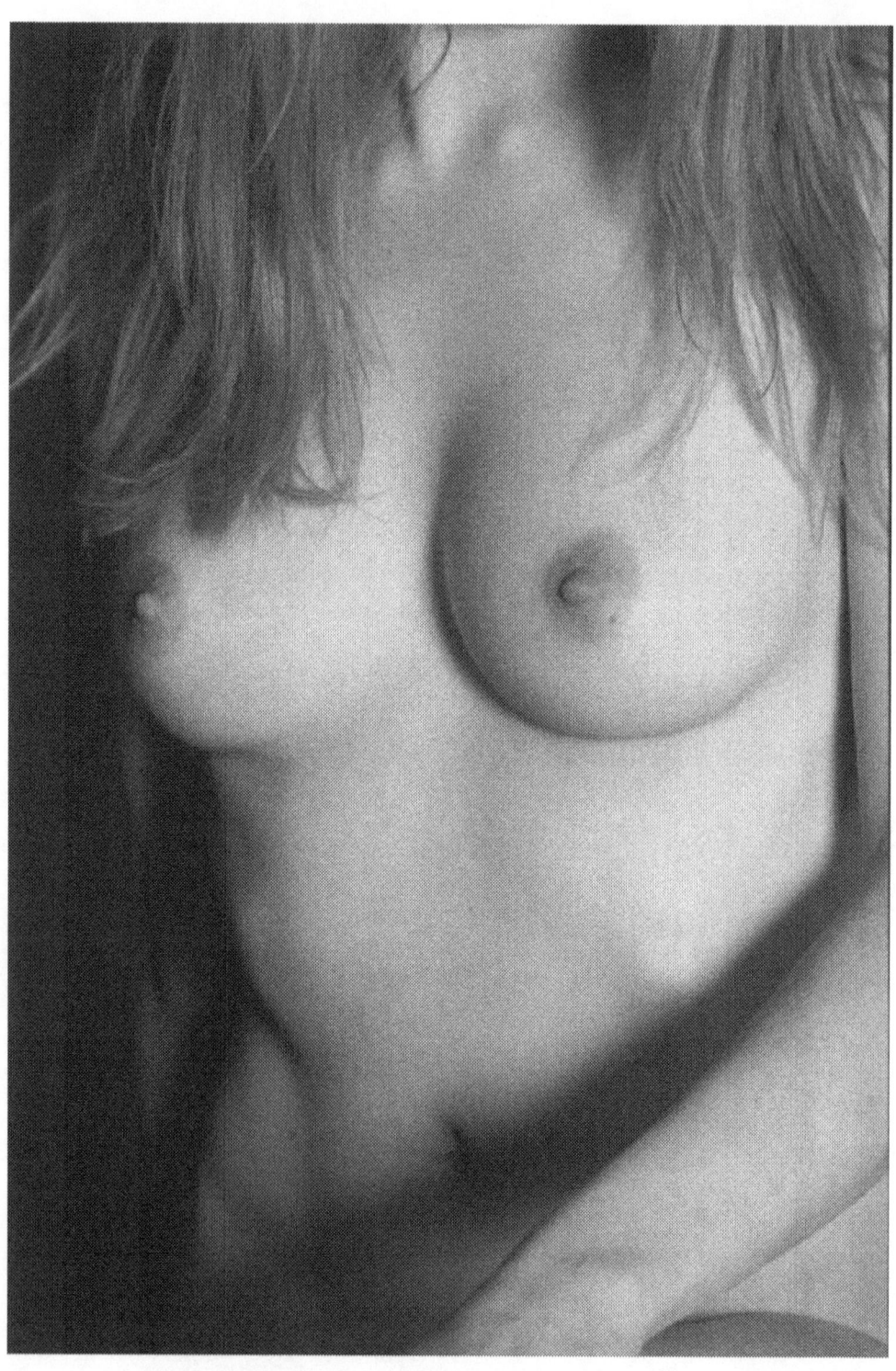

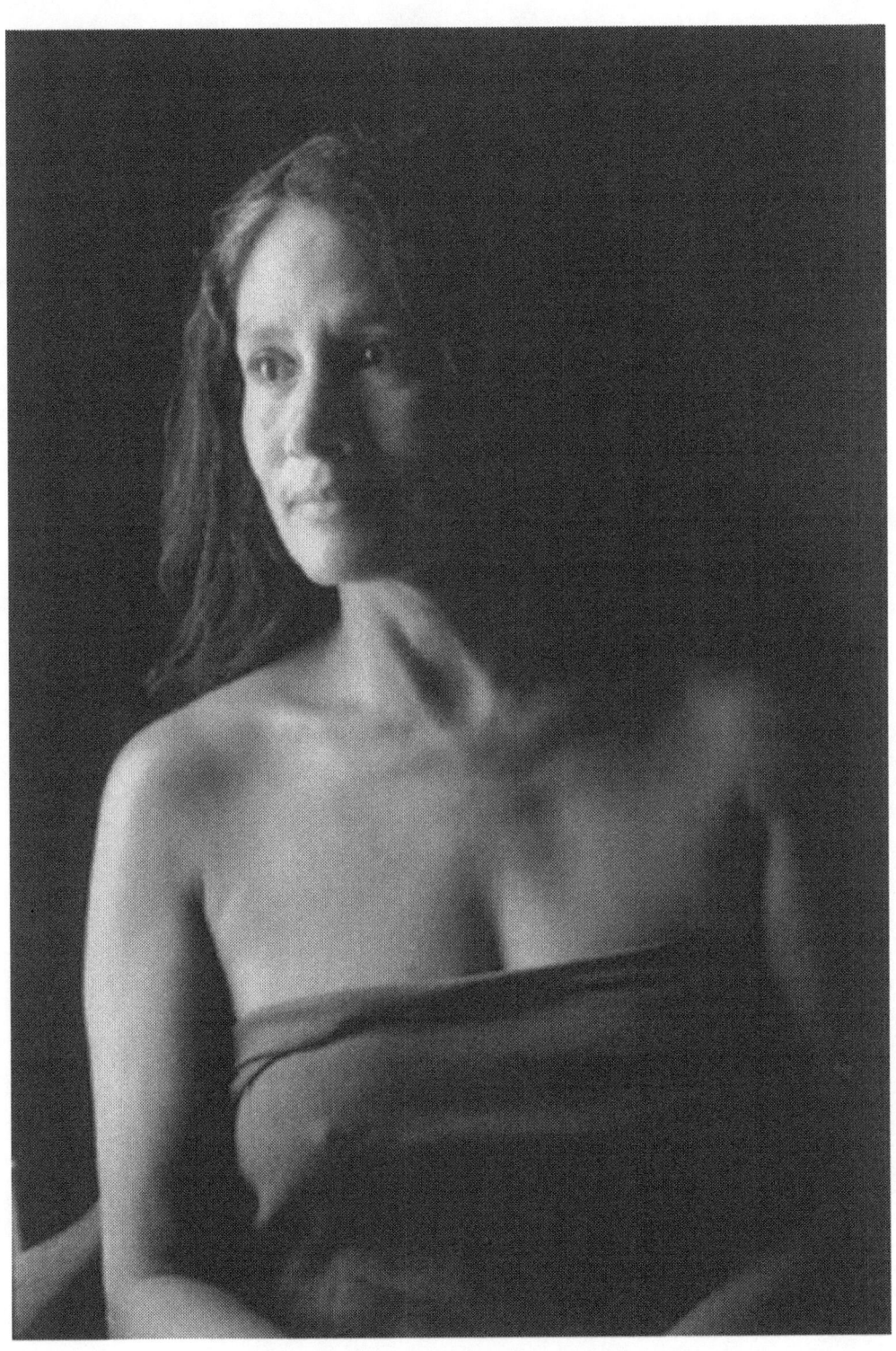

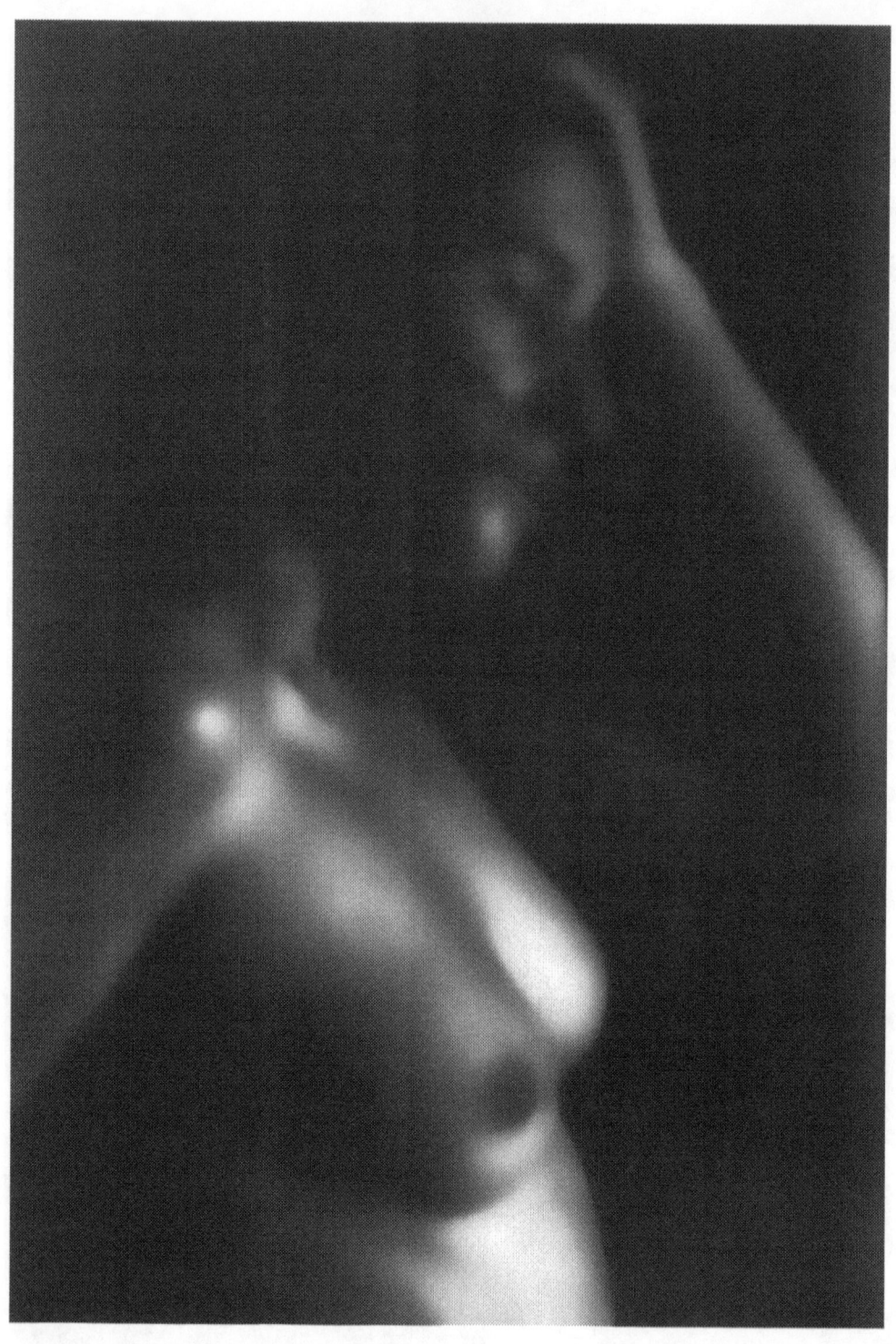

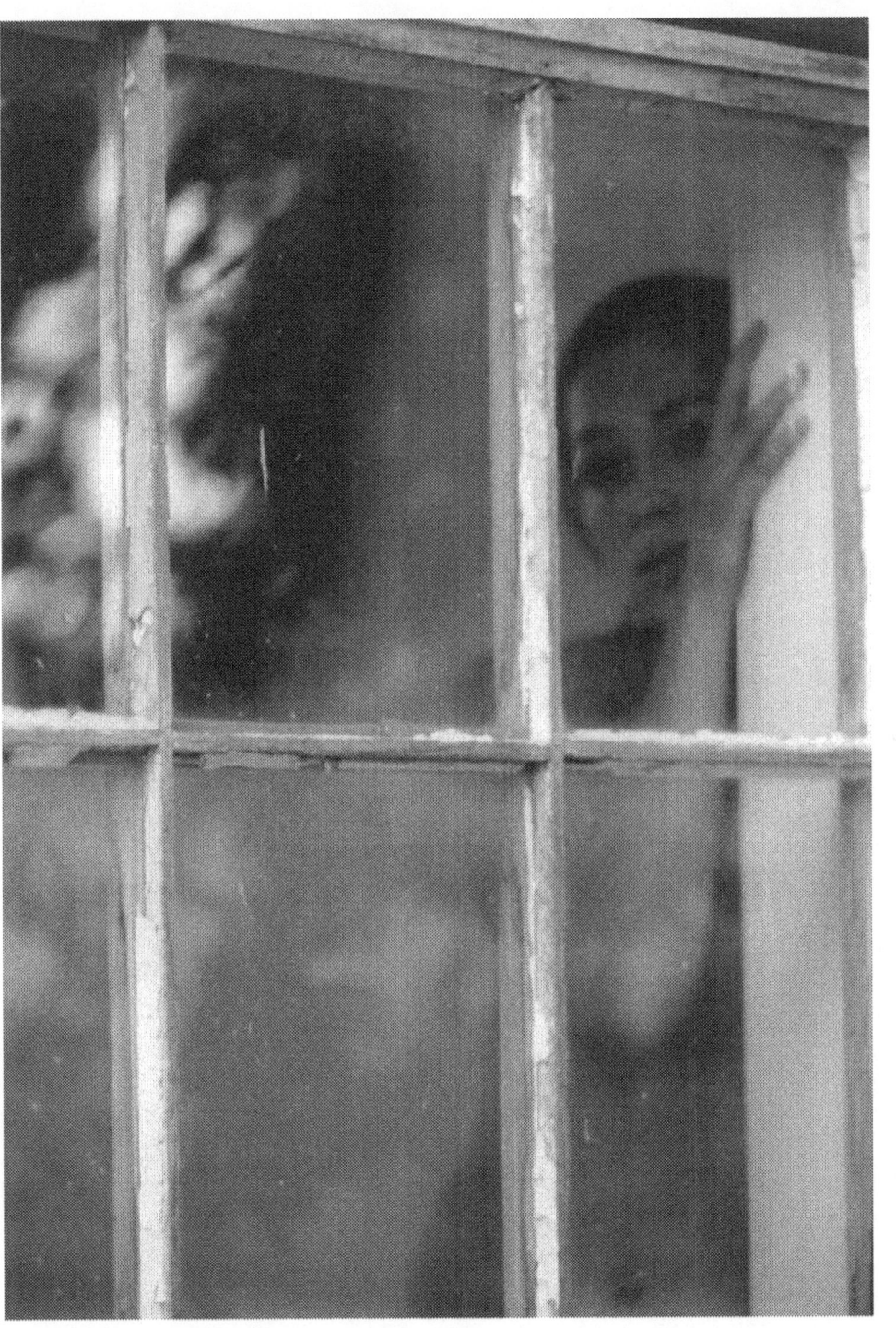

Mark I. Nelson

Most of my figure photography is done in my studio and I put a lot of advanced thought into this work. I think photography is 75% thought. Figure photography is my most personal work, coming from deep inside, in contrast to happening upon a landscape that captures my attention. I rarely use professional models—I prefer to work with people I know who have expressed an interest in working with me.

While serendipity certainly can play a part in creative work, working with a model requires that you have done your homework—you have another person's time (and maybe fees) to take into consideration. It is very embarrassing to start a shoot with a model and have no idea what you are going to do. The model doesn't just disrobe and something magic happens. An unprepared photographer also is somewhat unnerving to the model. So, I prepare in advance.

The ideas for my images usually center around current themes or concepts that are personally important to me. I keep a running list of ideas for images for each of these themes. I never know when I will get an idea, but when I do, I try to jot down a paragraph related to that idea as soon as I can. Sometimes, as I am writing down the idea, I will close my eyes and picture the image as if I am watching a movie. I let the movie play itself out and try to write down what the image looks like at the end of this movie, including composition, lighting, setting, the feeling of the image, and how the image fits into the theme category.

Before a shoot, I pick which image ideas I would like to work on with a specific model. Experience usually tells me which ideas seem to fit the personality of a particular model. I discuss these ideas with the model and we narrow the field to those that seem to intrigue the model most. By involving the model this way, the model becomes invested in my work and a wonderful partnership develops, leading to rich imagery.

General Poses

This section of poses can be used as a starting point for both artists and models. Some of the poses on the following pages are commonly used, but they are presented here as a launchpad, a place to begin.

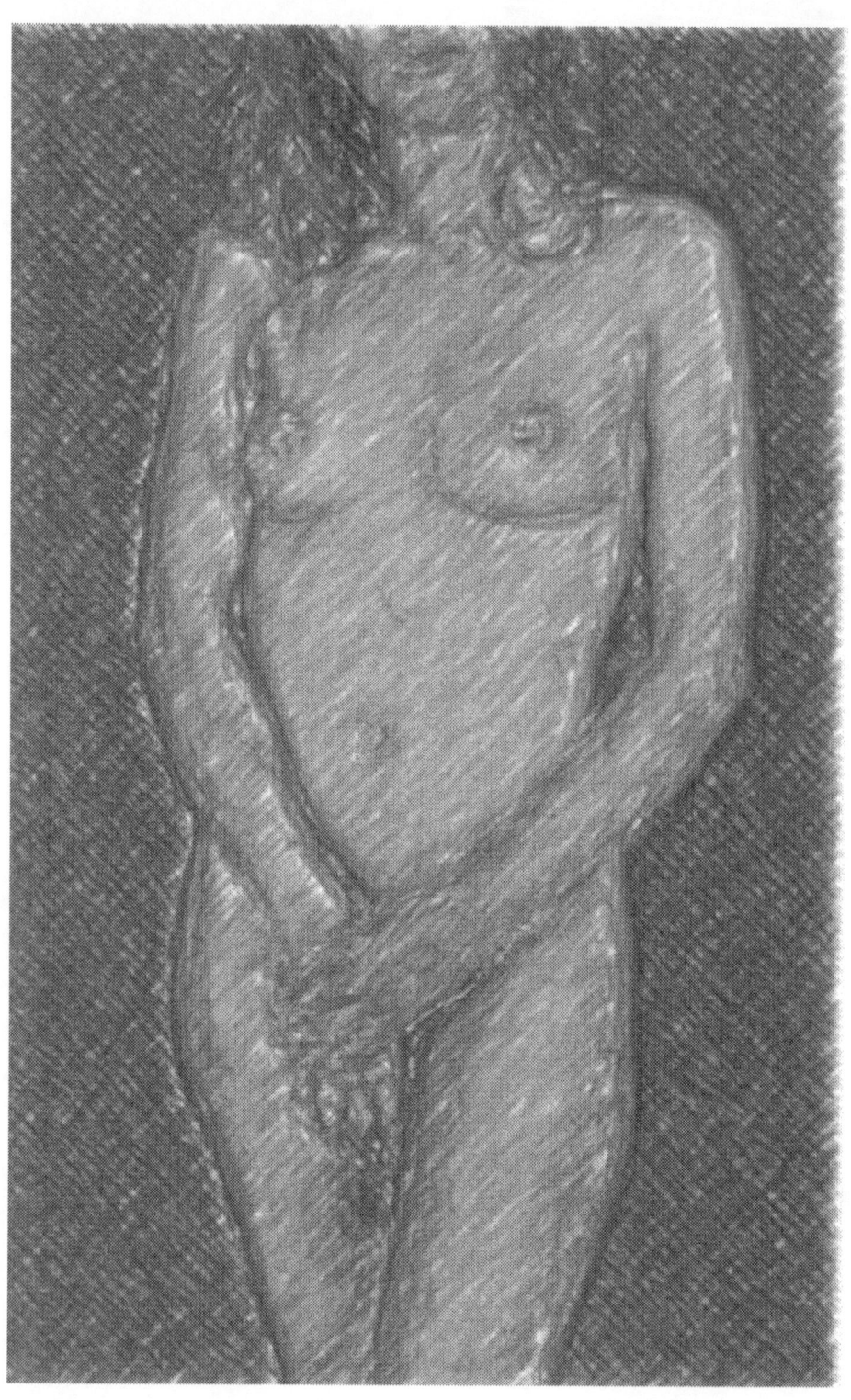

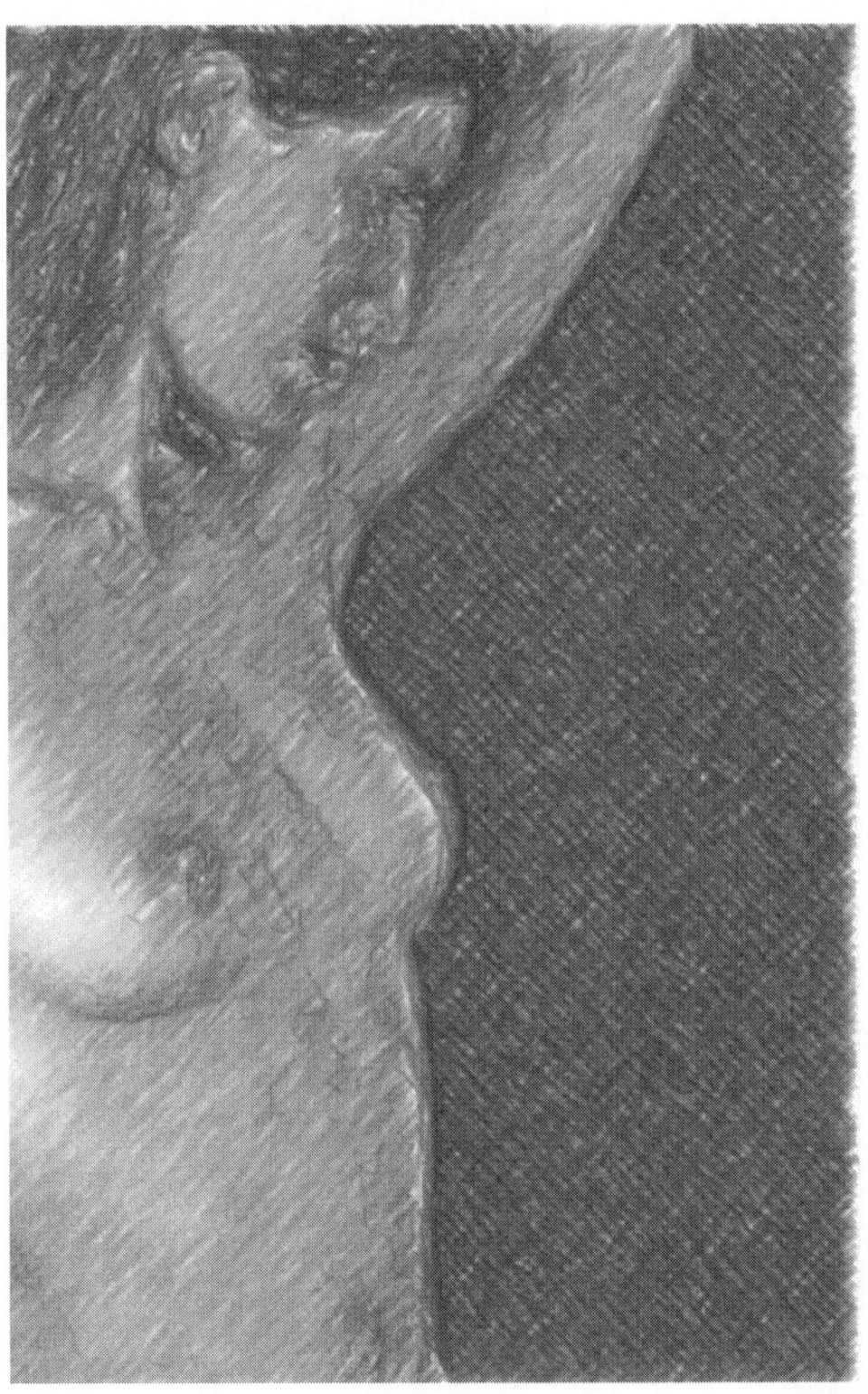

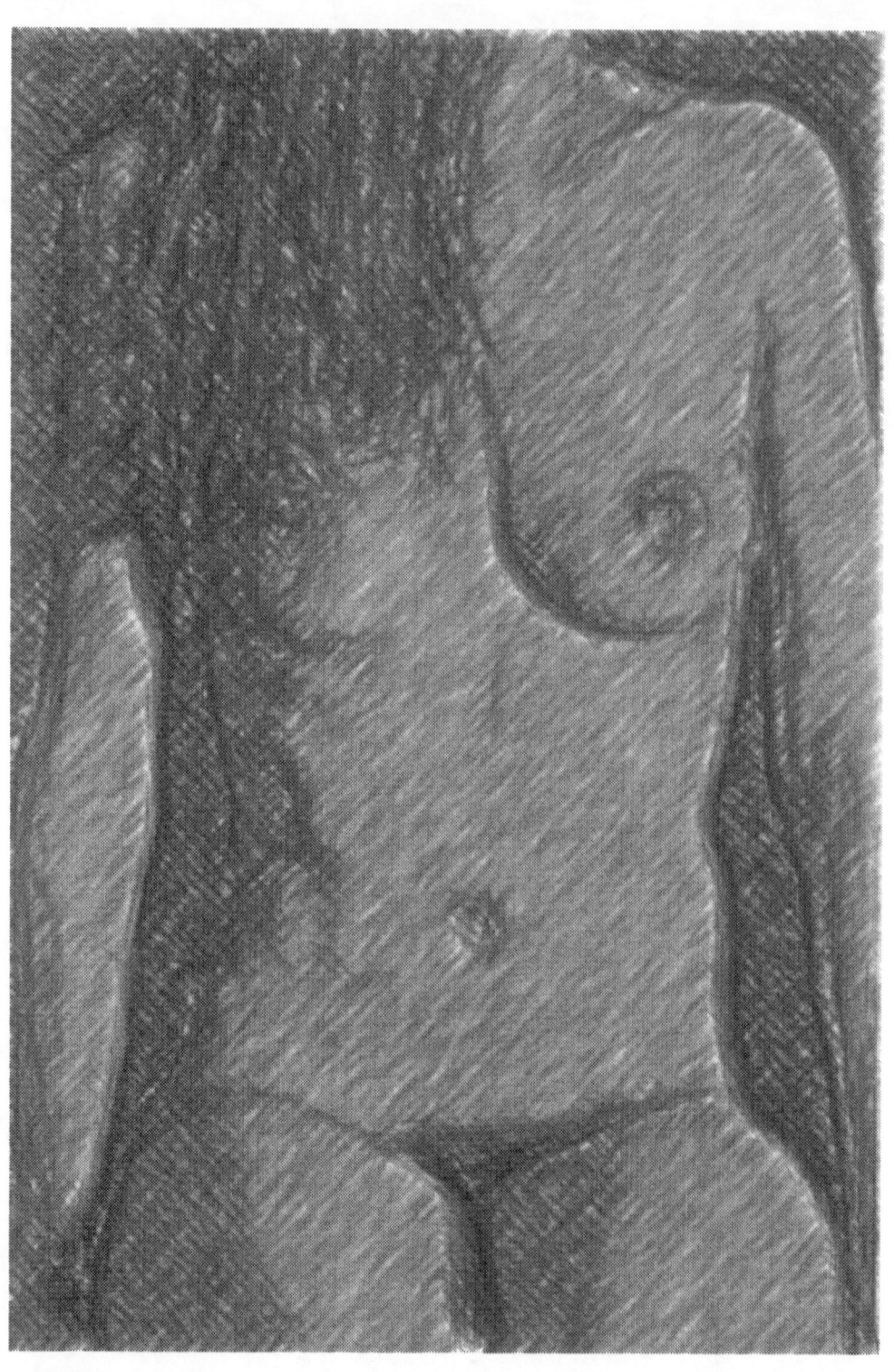

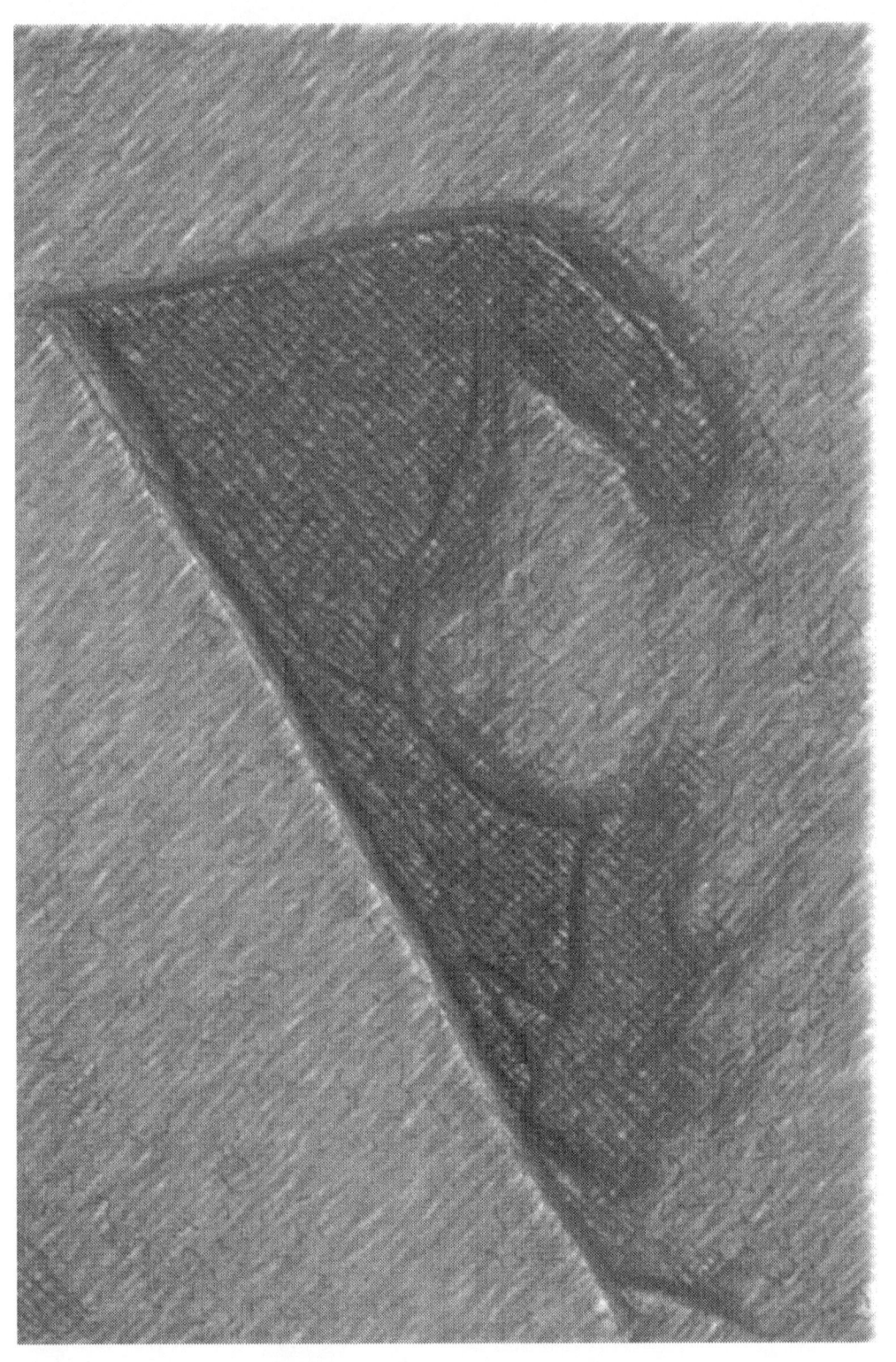

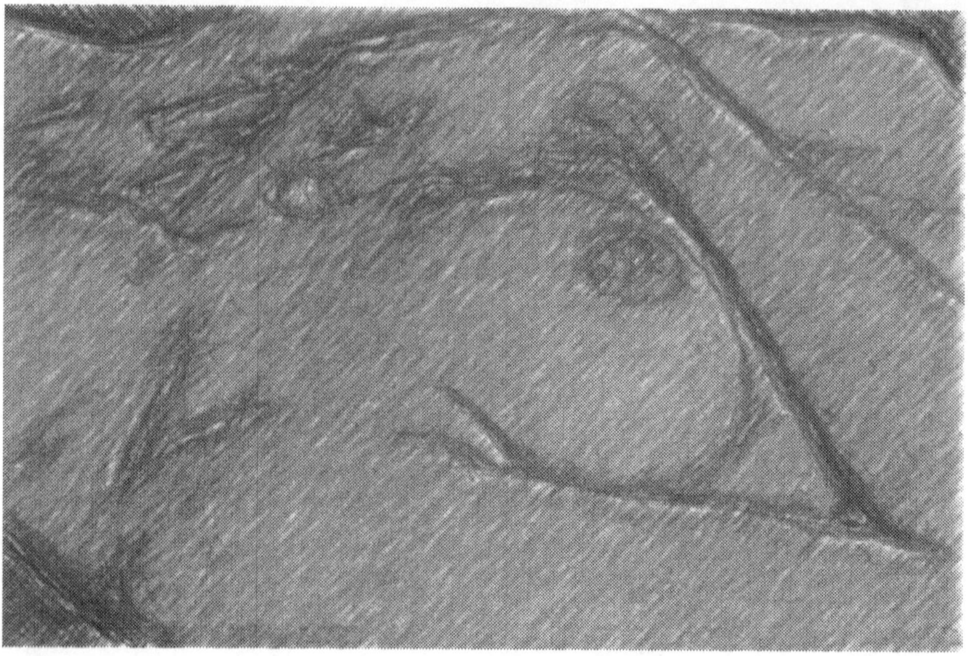

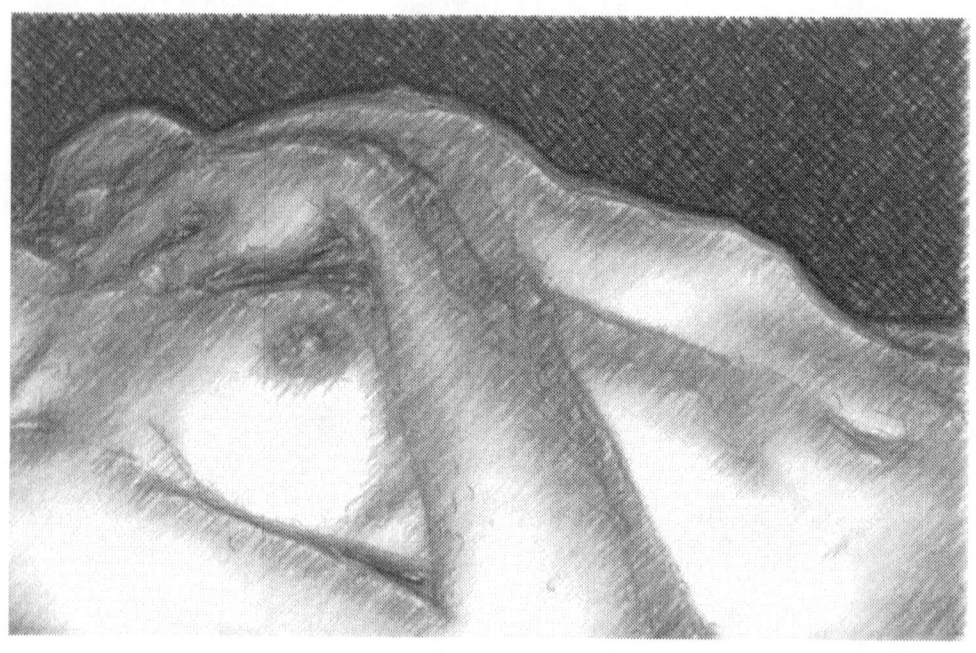

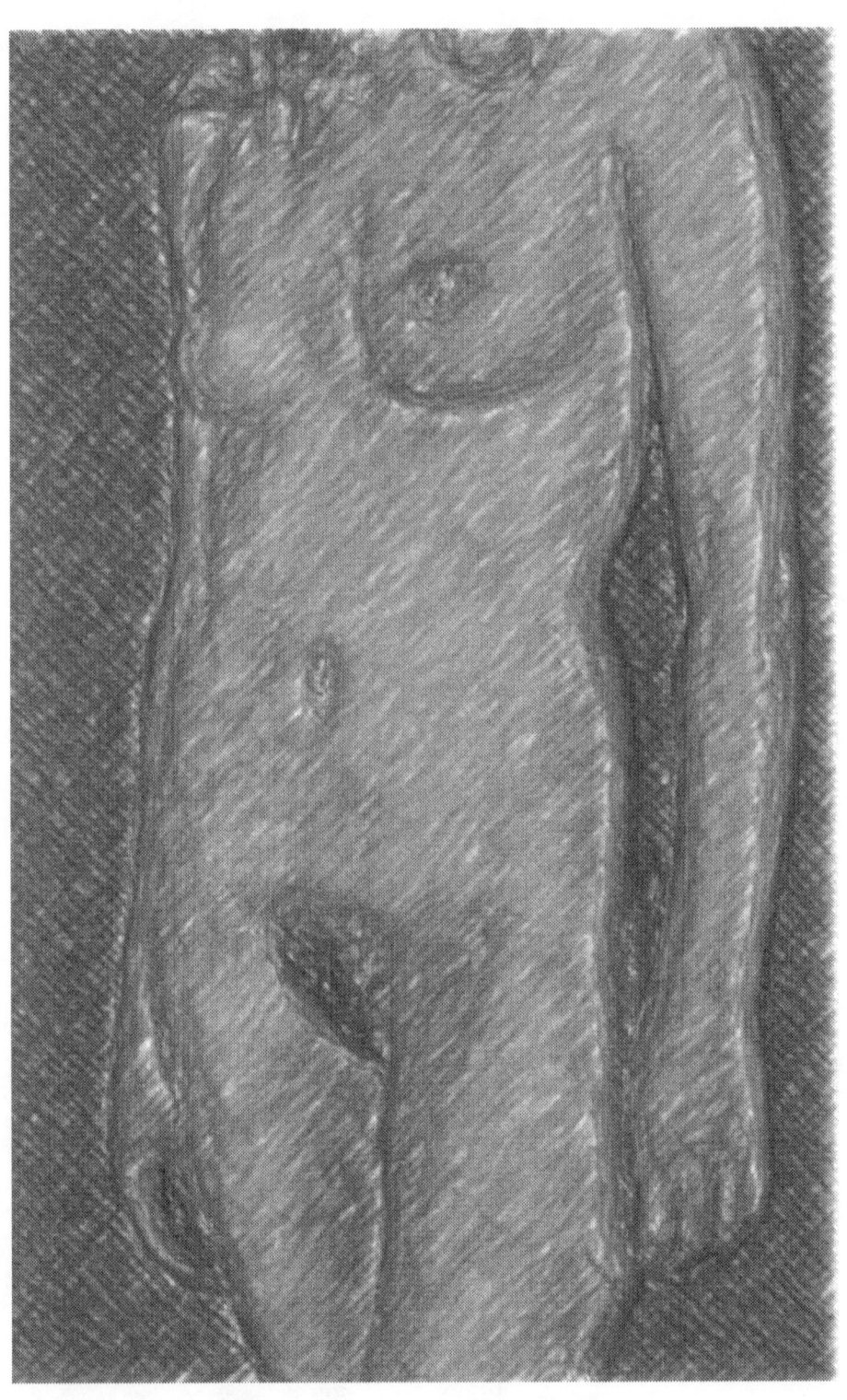

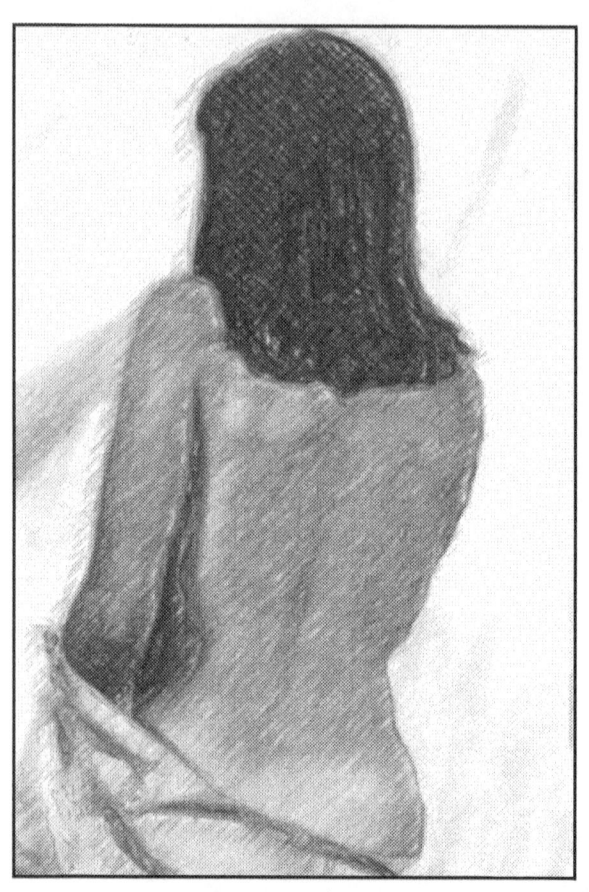

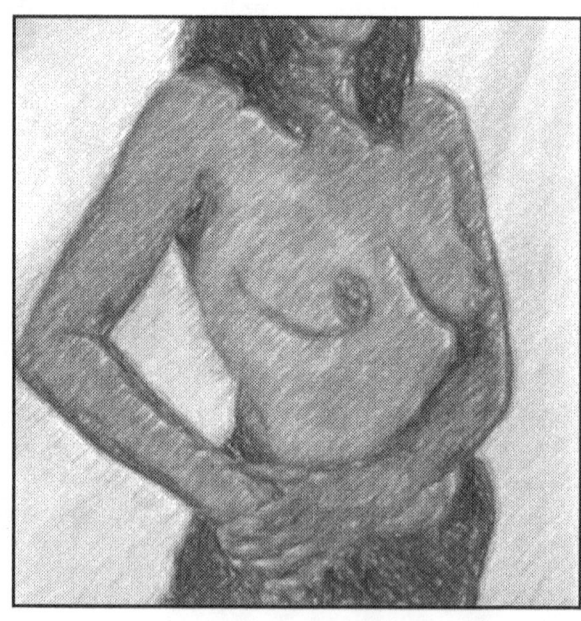

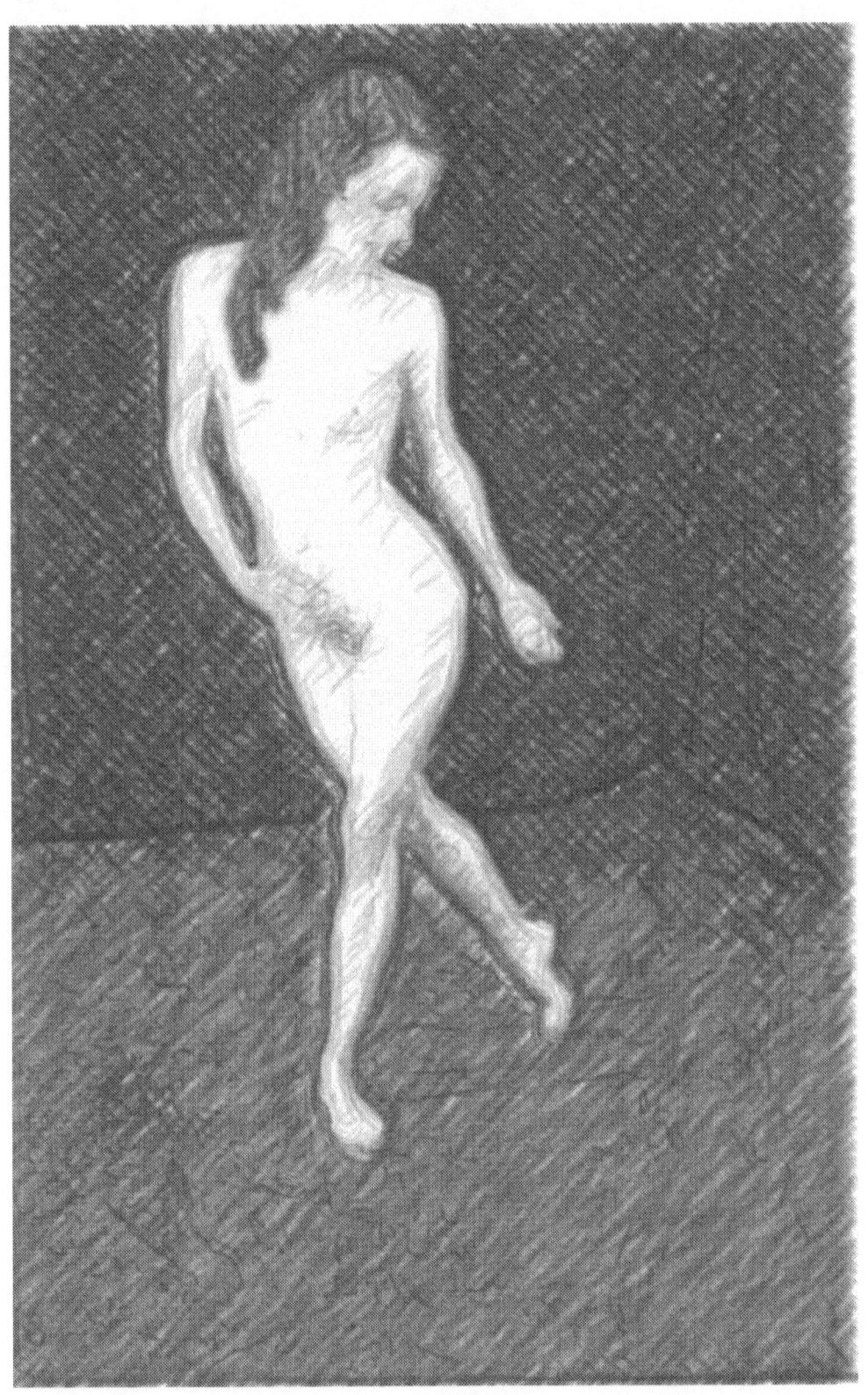

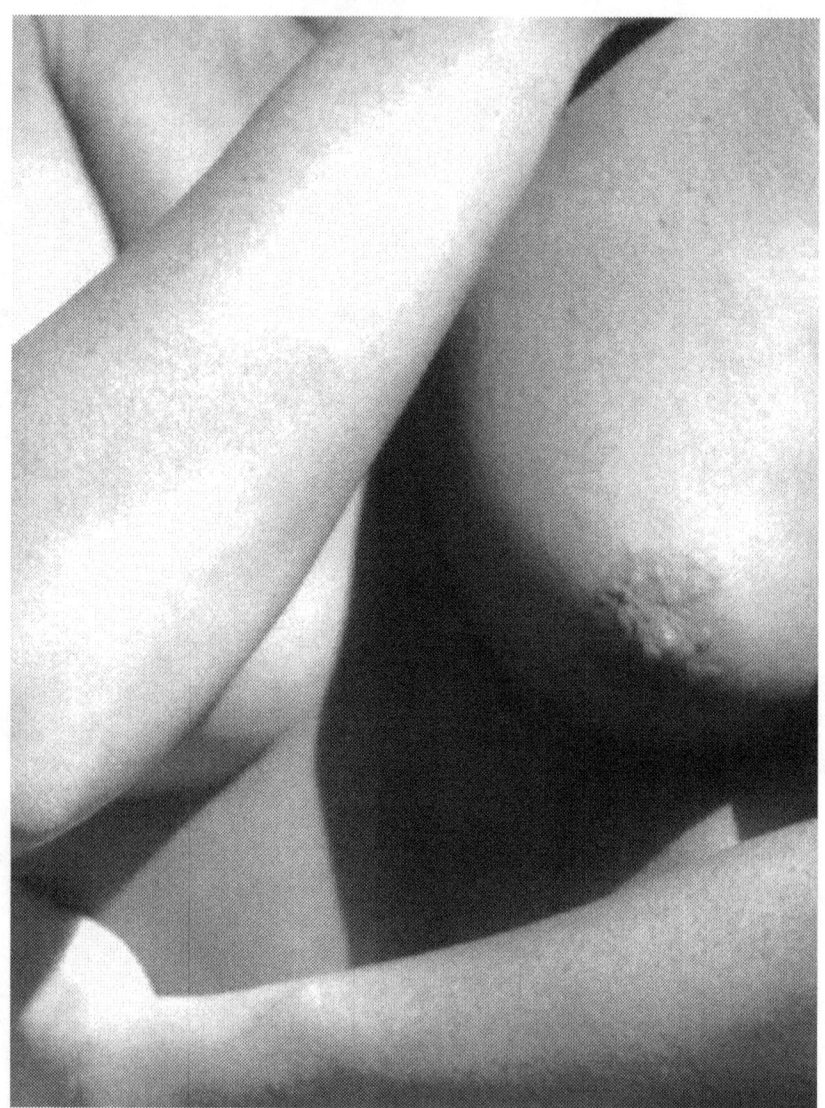

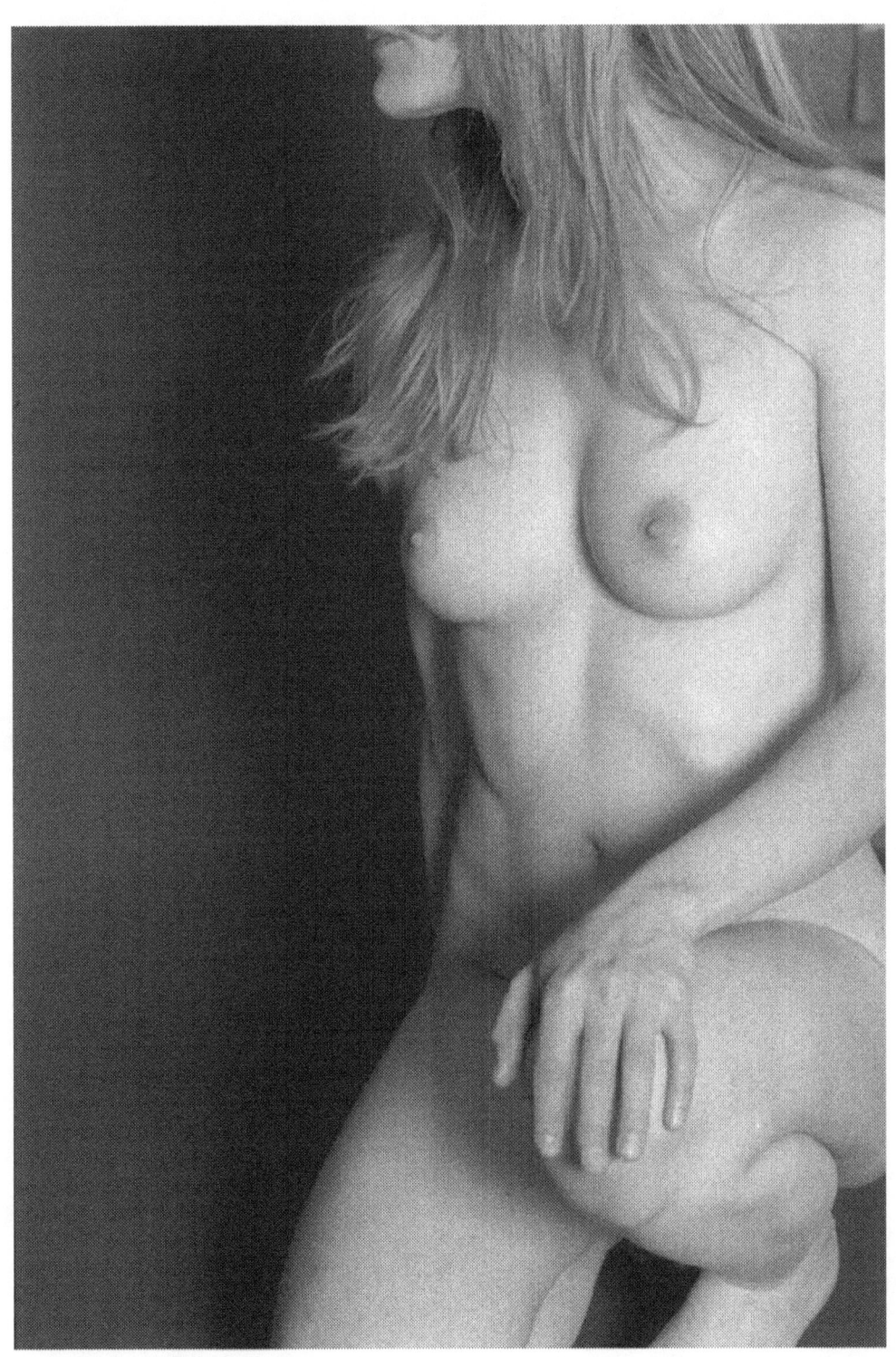

Kim Weston

Most of the time when I work with a model, I have scripted what we will be photographing that day. This could be as elaborate as a storyboard with each individual photograph mapped out, each choreographed as I have seen it in my mind's eye. Or it could simply be an idea I wish to convey no matter what the circumstance.

I have always found it paramount that model and photographer know what the finished piece should be. The model is not just a plug in naked person; they are an integral part of the photograph. Their personality, their movements all come together with the photograph to make a whole. I find that I need to photograph a model at least three times before a trust and willingness to continue can be achieved.

Without this bilateral agreement the model is just a thing and the photographer is a wanting pursuer of only flesh and light.

Brigitte Carnochan

My models come from my circle of friends and fellow students in dance classes. I prefer models who pose from the heart and from the natural inclinations of their own bodies. I often use myself as a model. I'm available at the right times, and I know what the photographer is looking for.

Some friends have modeled for me often—we work together well and I can demonstrate a possible pose, add a prop (a piece of fabric or a flower) or suggest a feeling as a starting point for movements that lead to poses, a process they've grown to understand and usually improve on.

I always watch the model for that informal moment when she feels comfortable and at ease—even if it's in between specific poses. That will often be the perfect moment of composition; form, feeling, and light, all falling into a harmony I couldn't have predicted: the surprise and satisfaction of collaboration.

Outdoors

When shooting outdoors, many factors can come into play, such as the forces of nature or the intensity of the light. On more than one occasion, a good shot has been disrupted by a sudden appearance of gale-force winds, or an unexpected summer shower. Nevertheless, shooting the figure in a forest surrounded by the epitome of nature itself can be awe (and image) inspiring.

Being in a quiet location can also be very relaxing. In a few instances I have had to nudge a reclining model awake because she has become too comfortable.

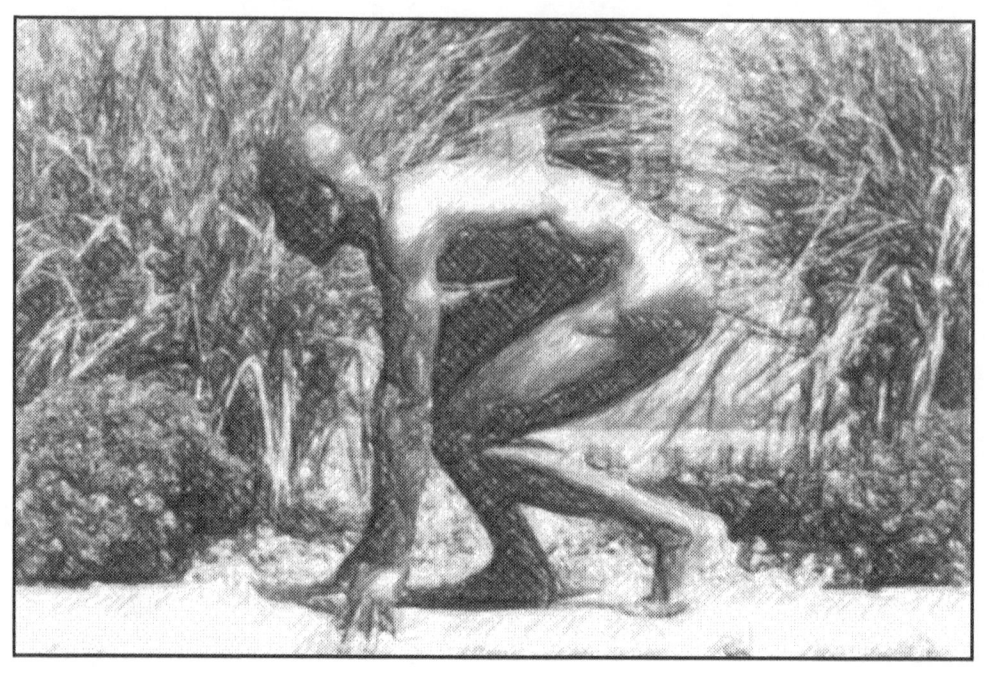

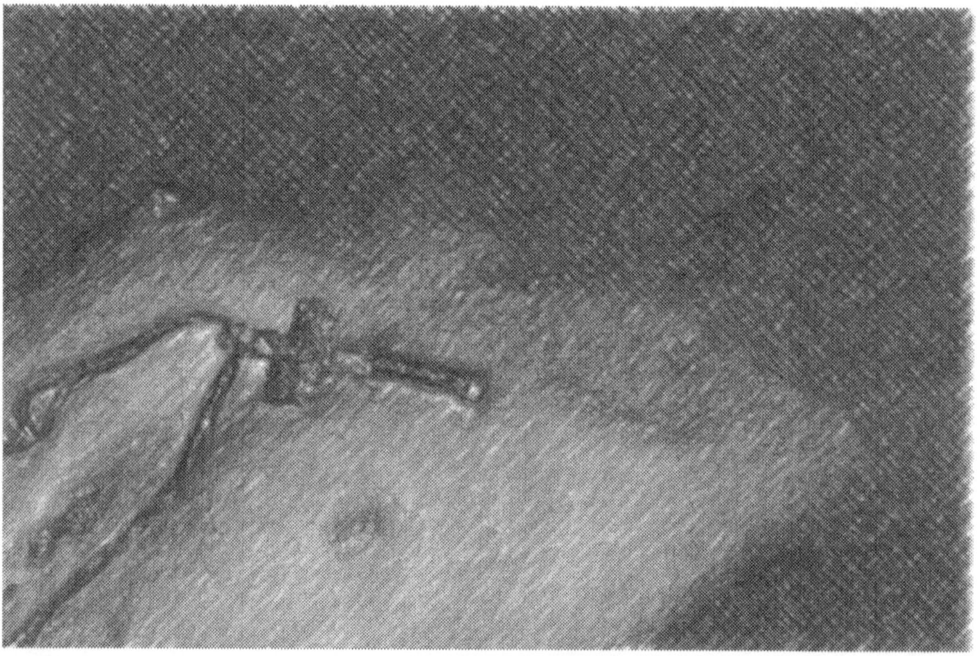

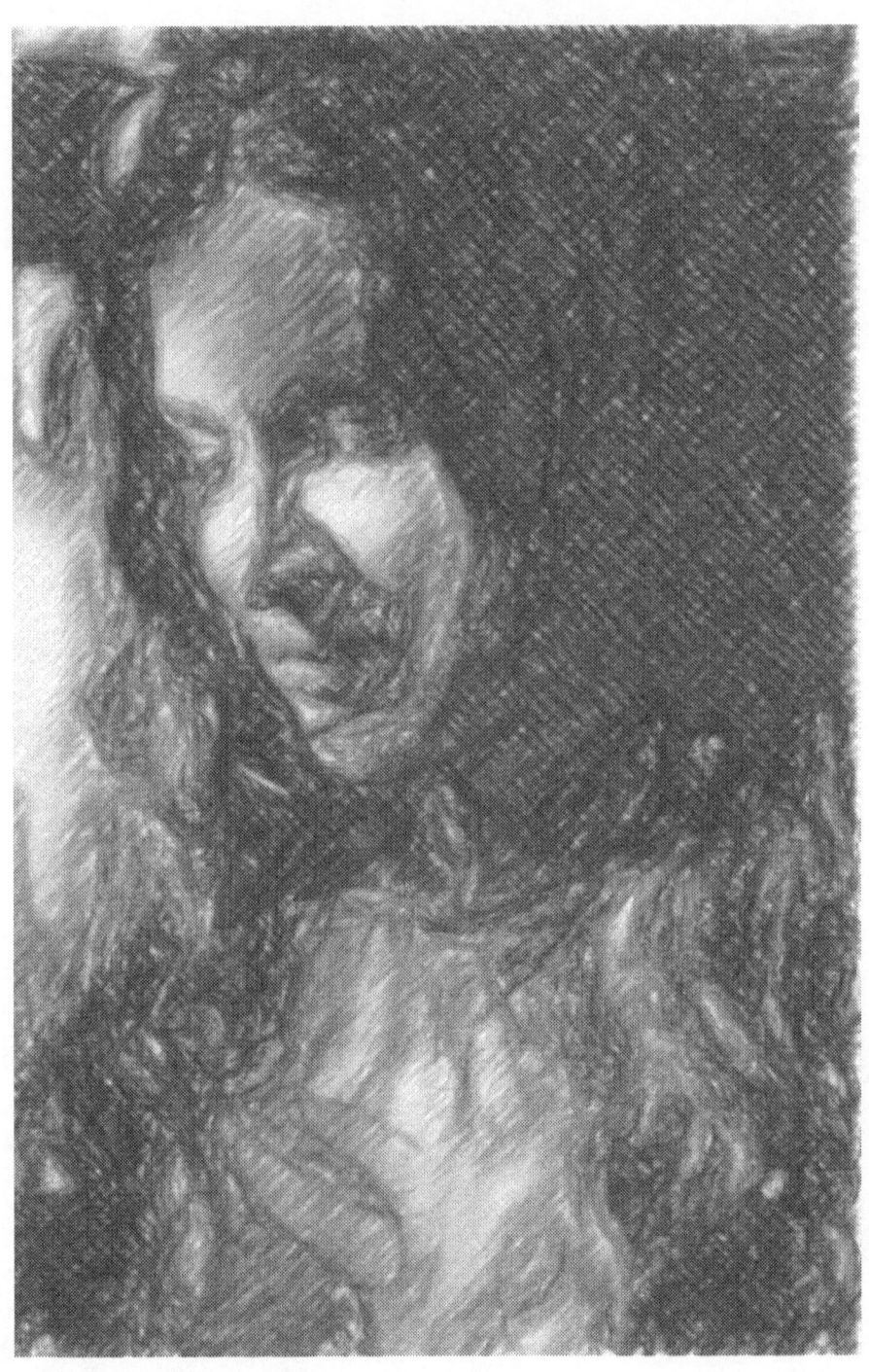

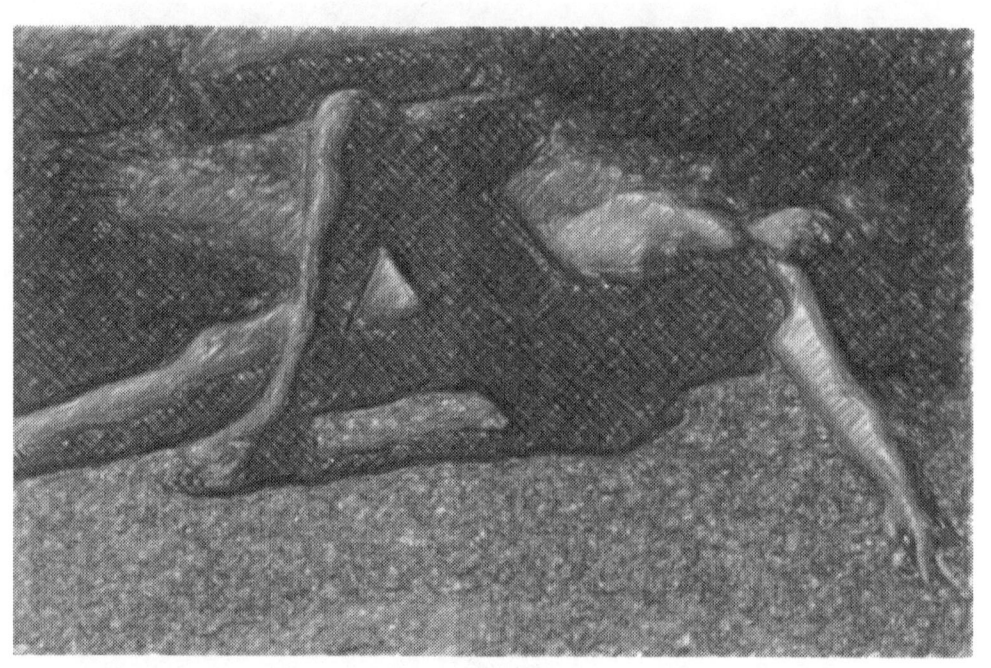

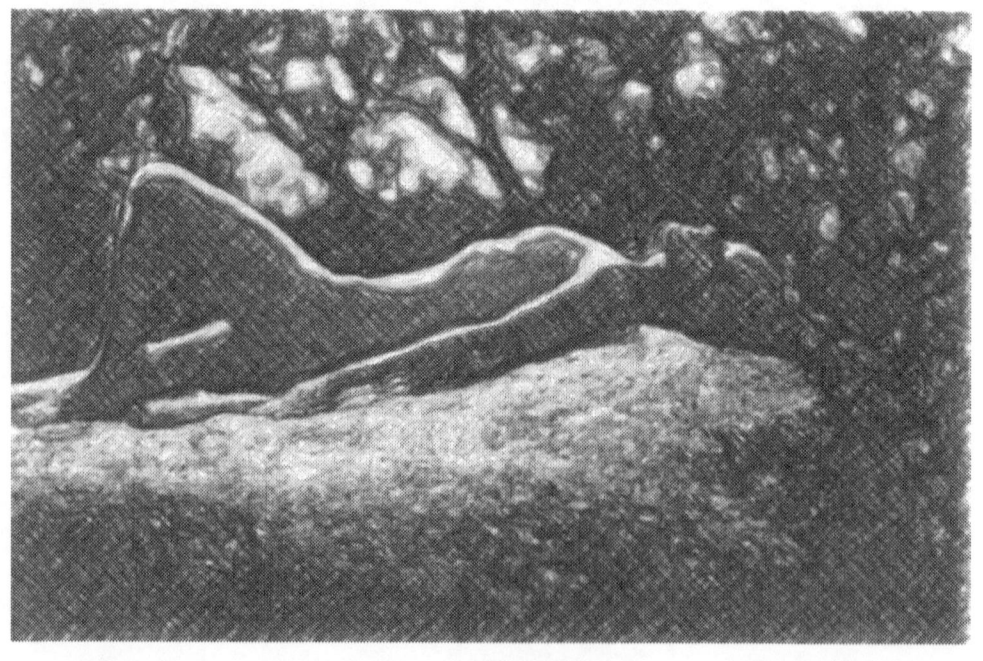

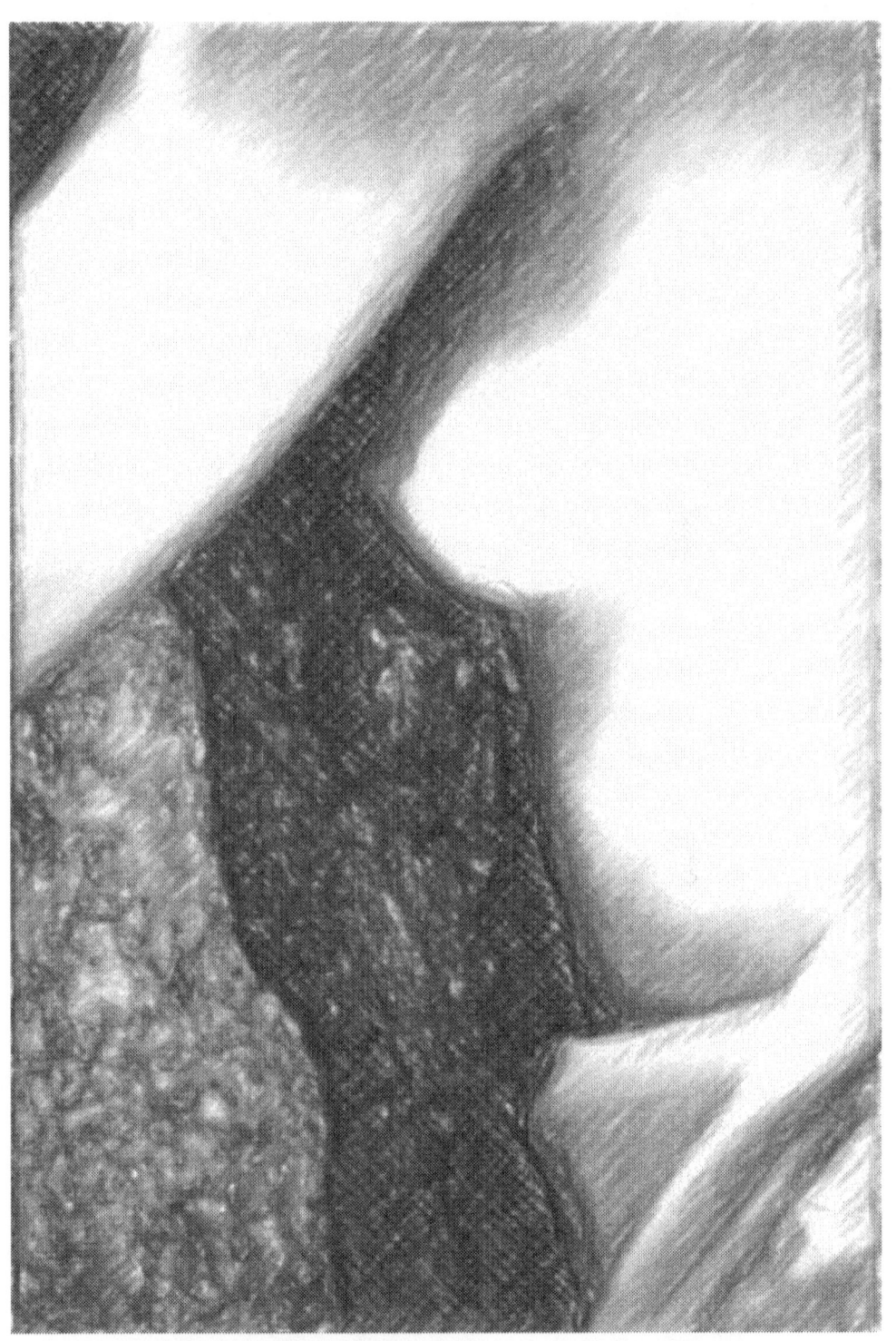

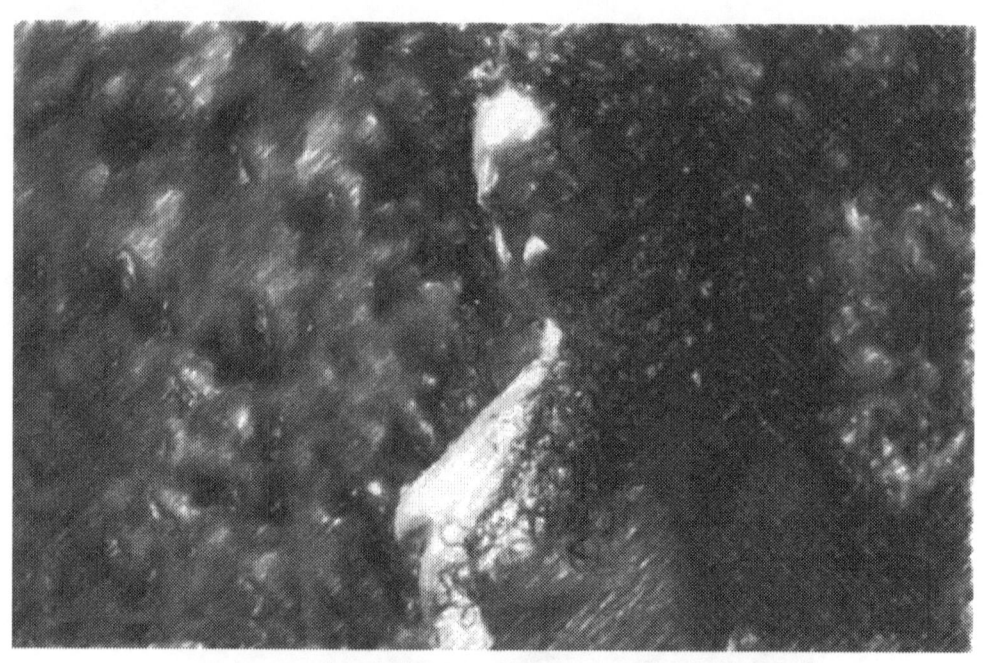

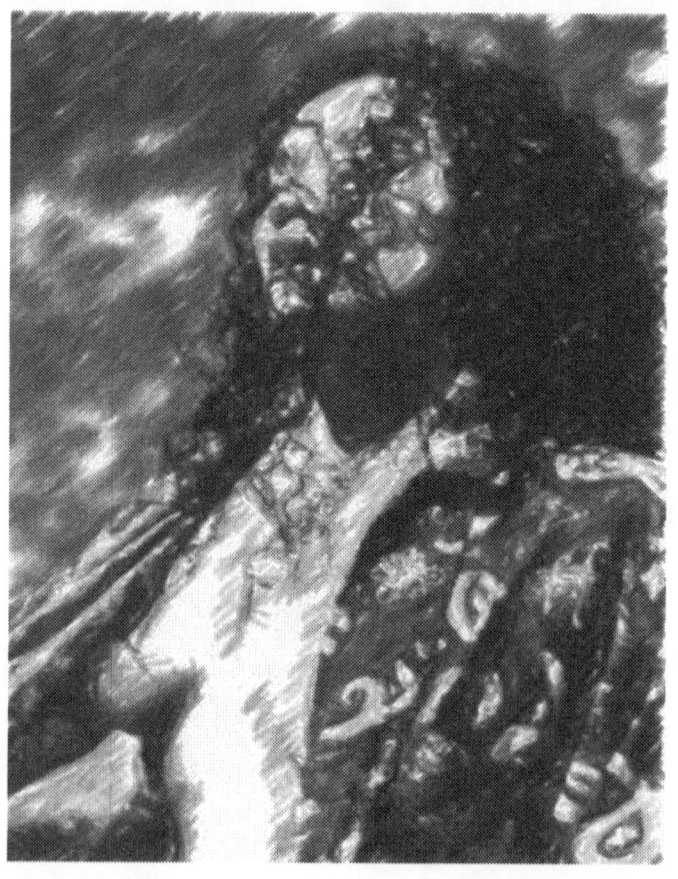

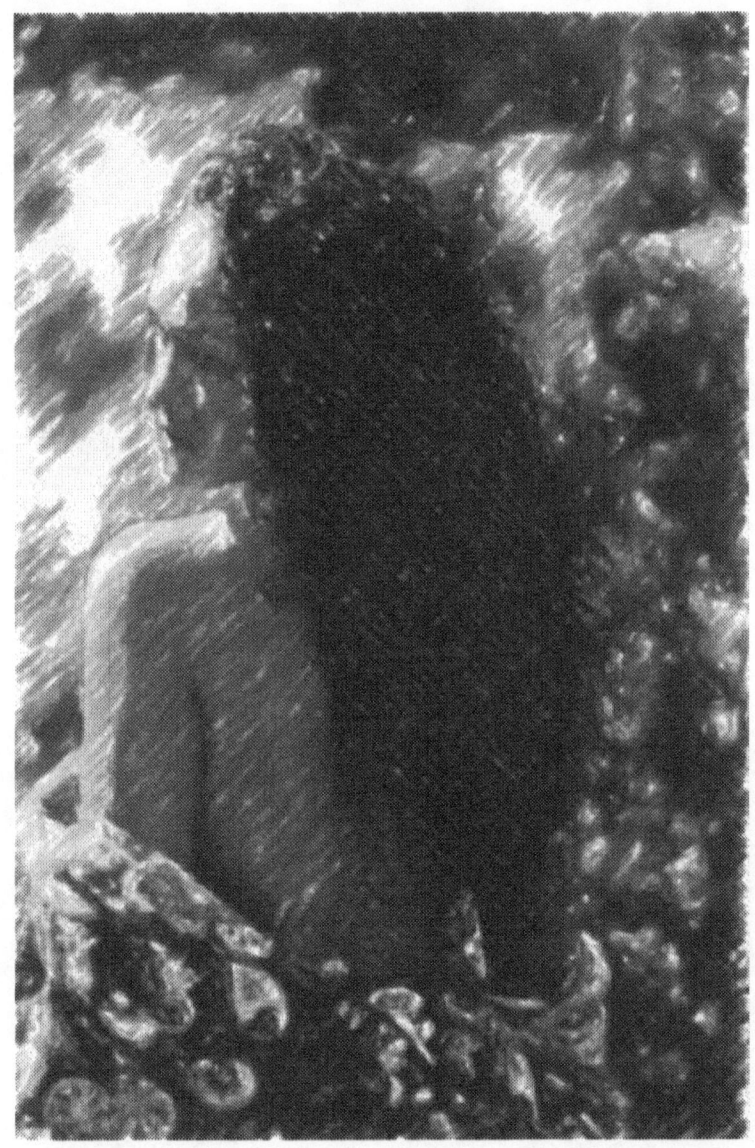

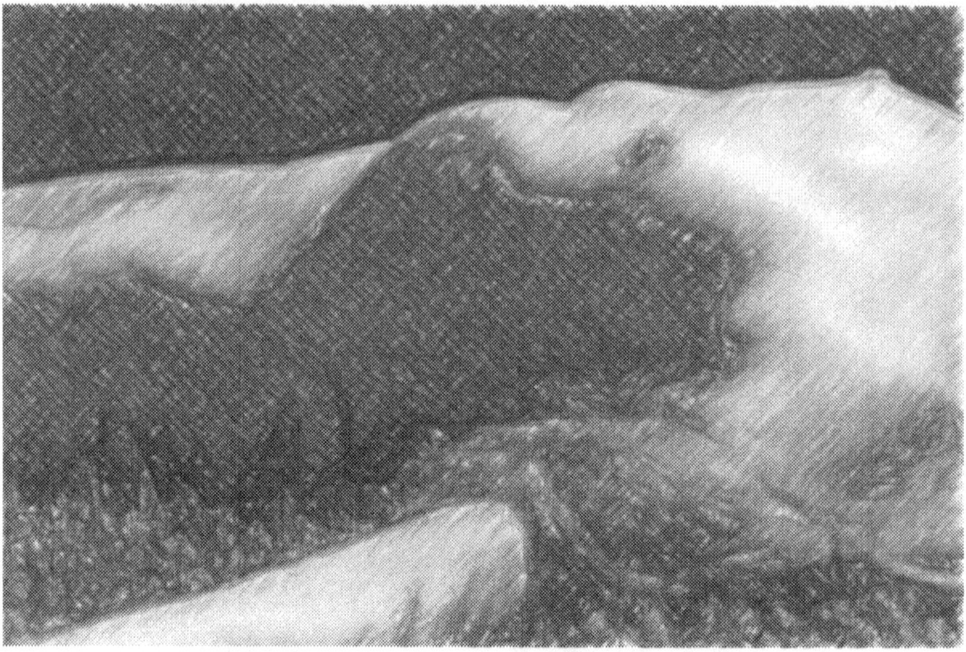

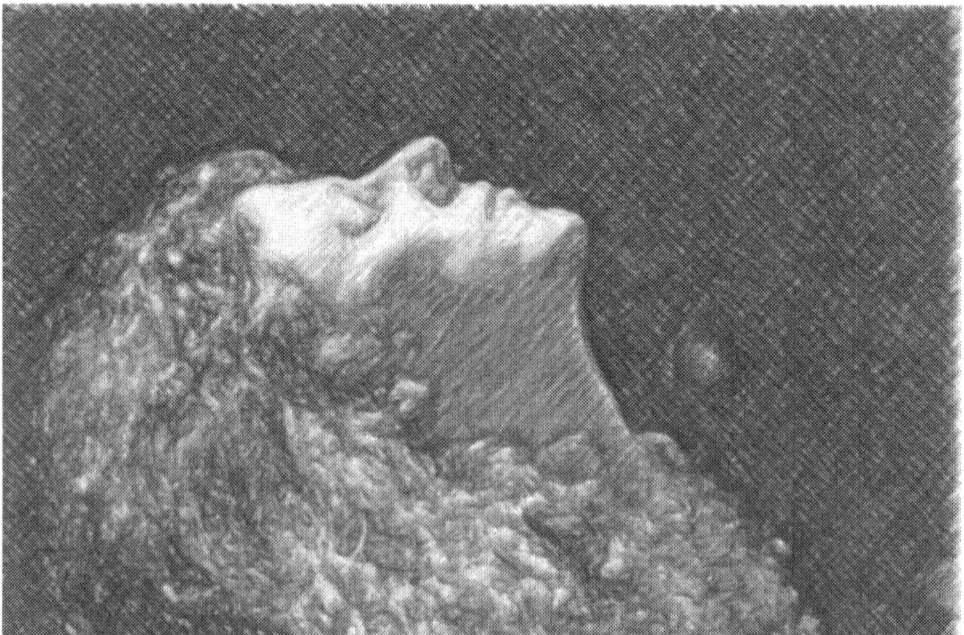

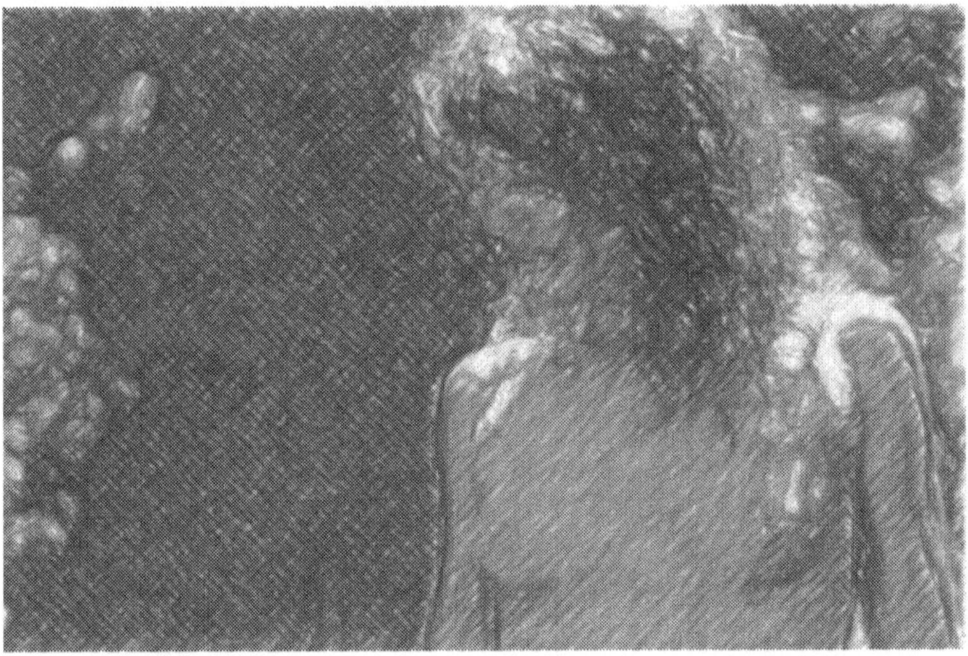

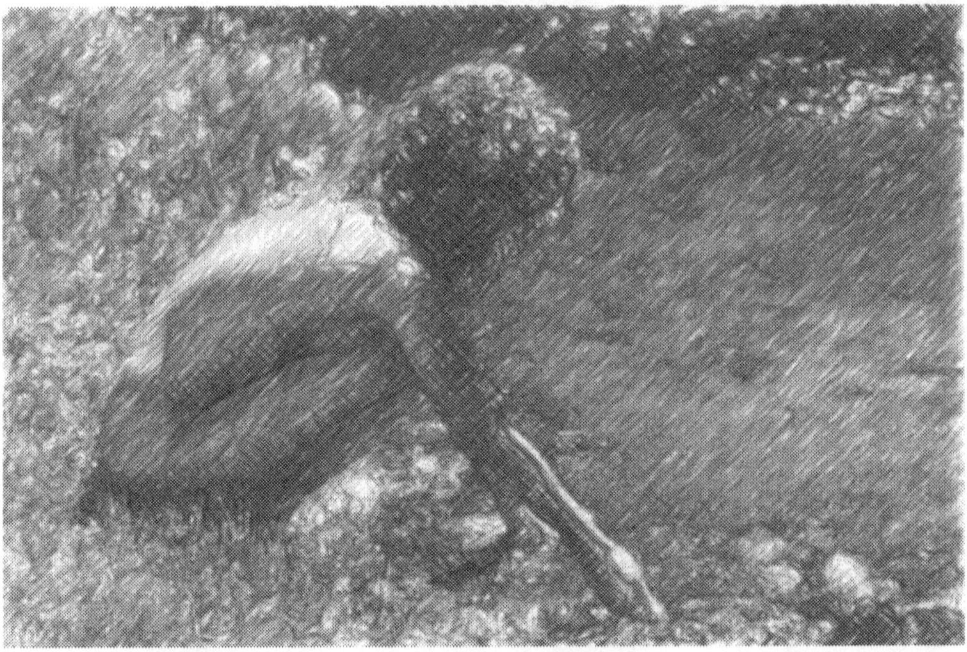

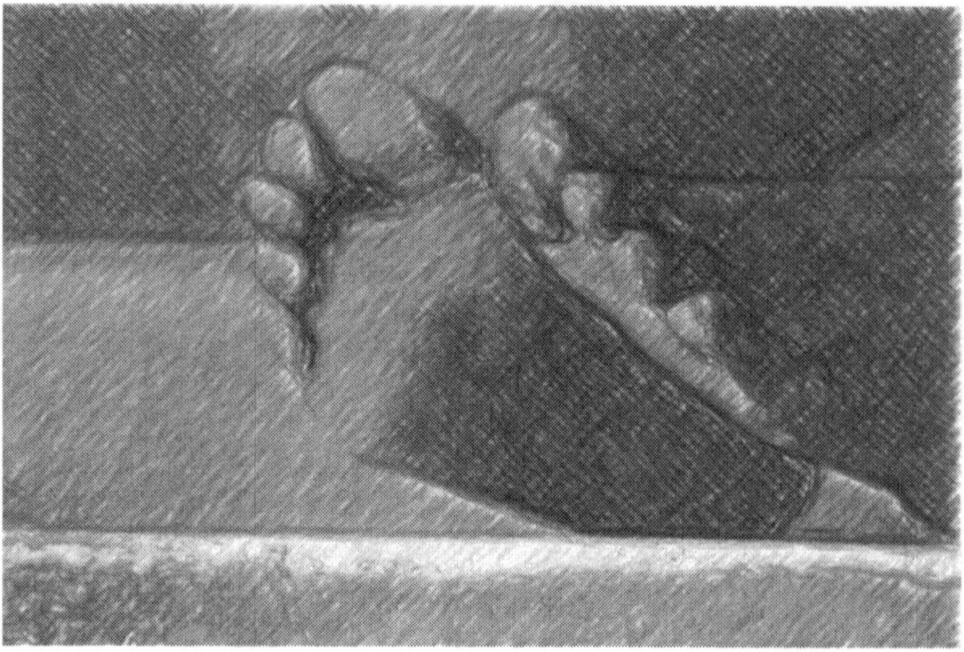

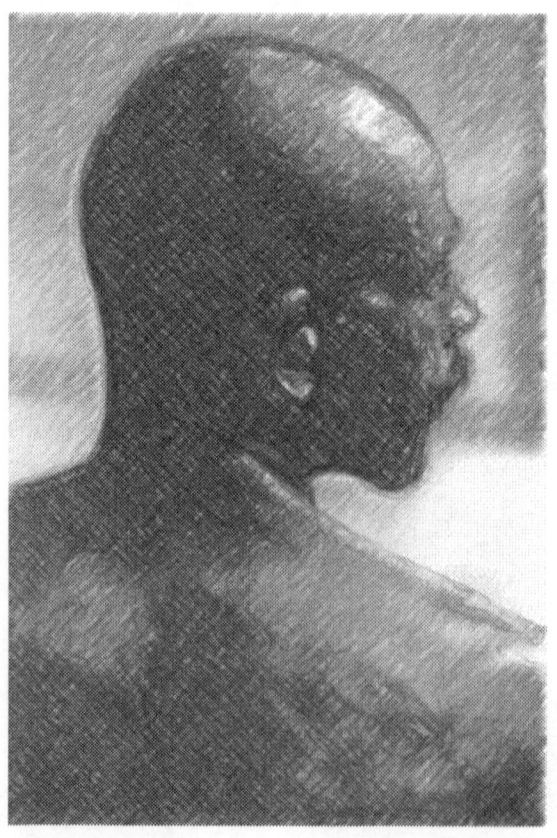

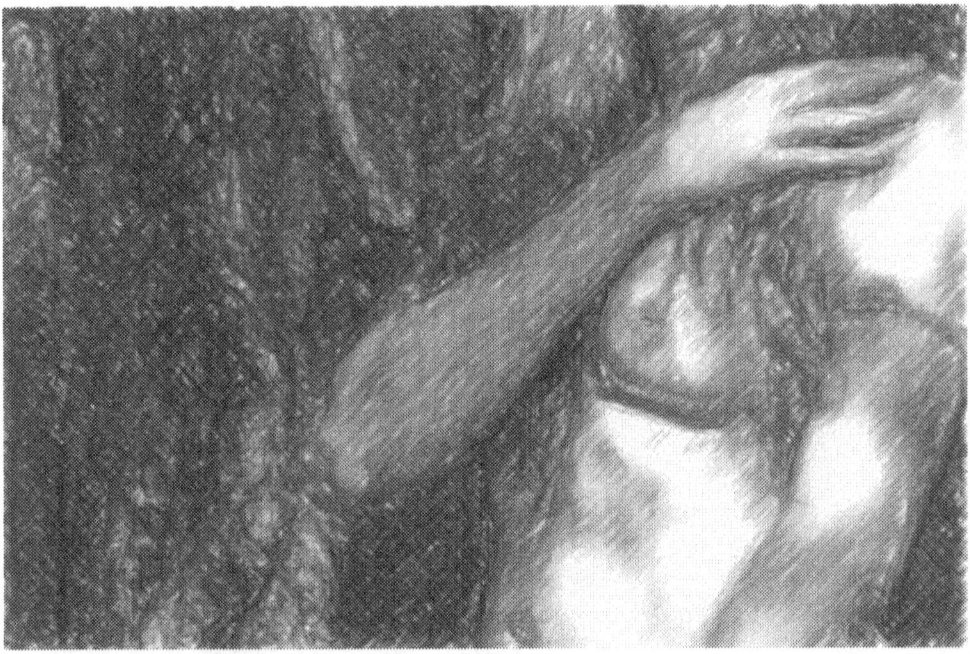

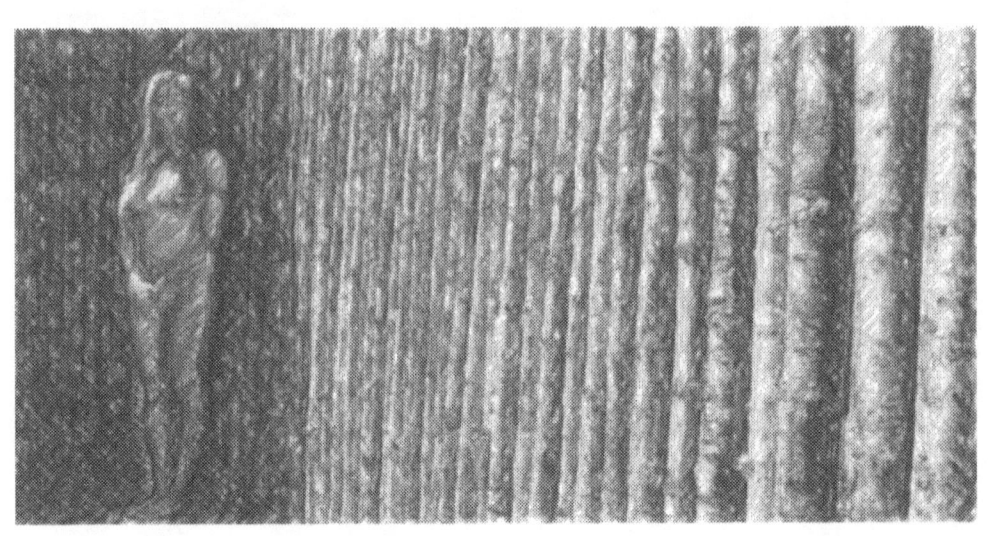

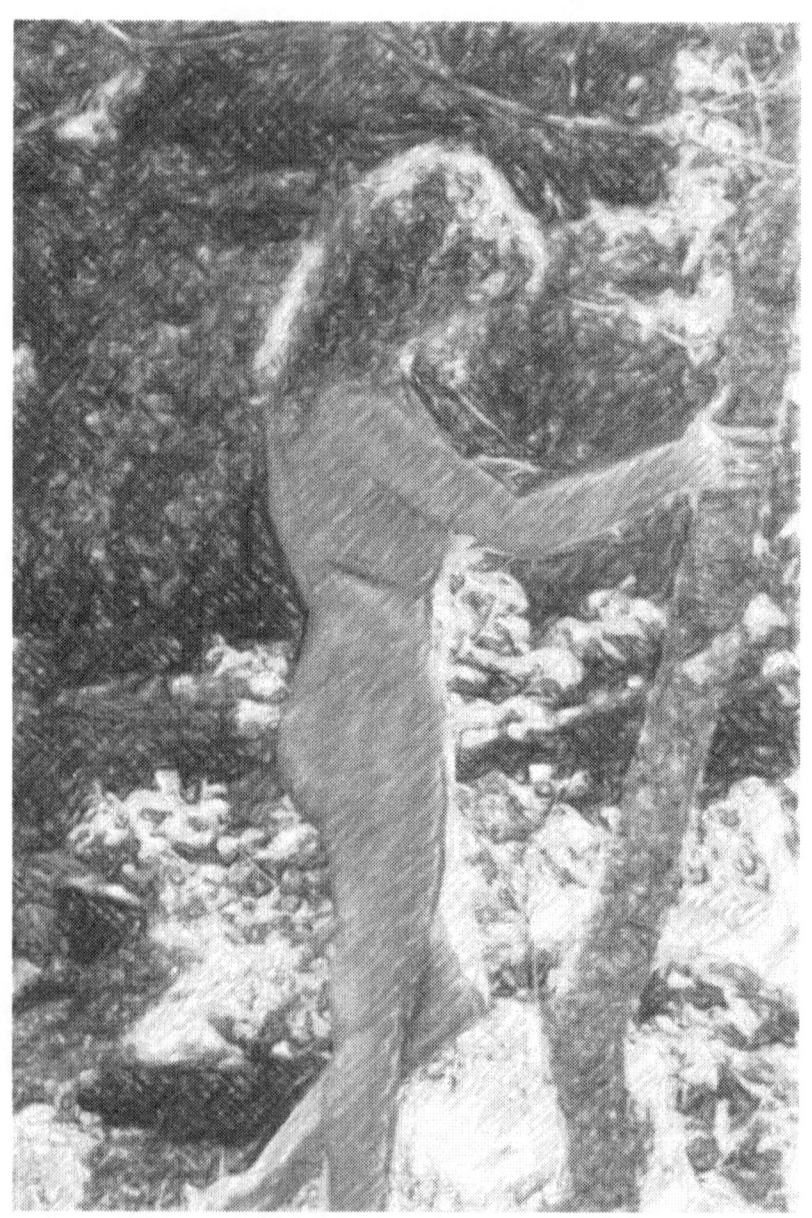

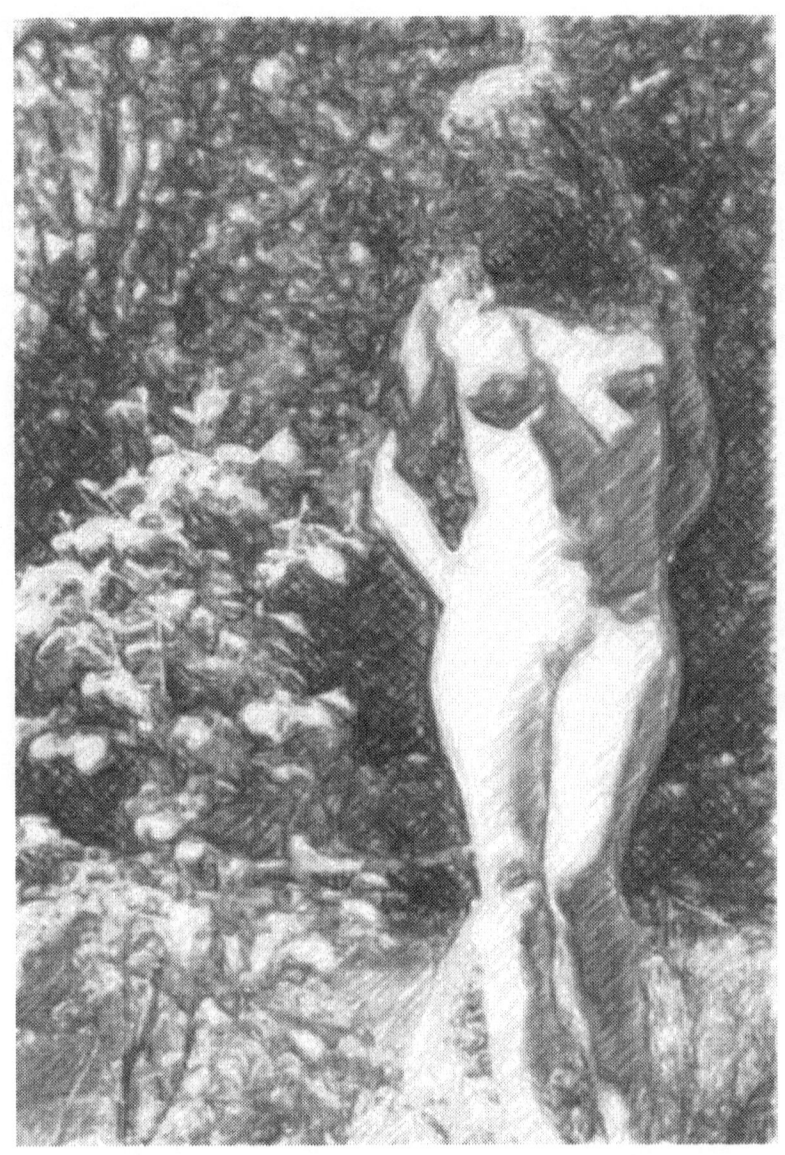

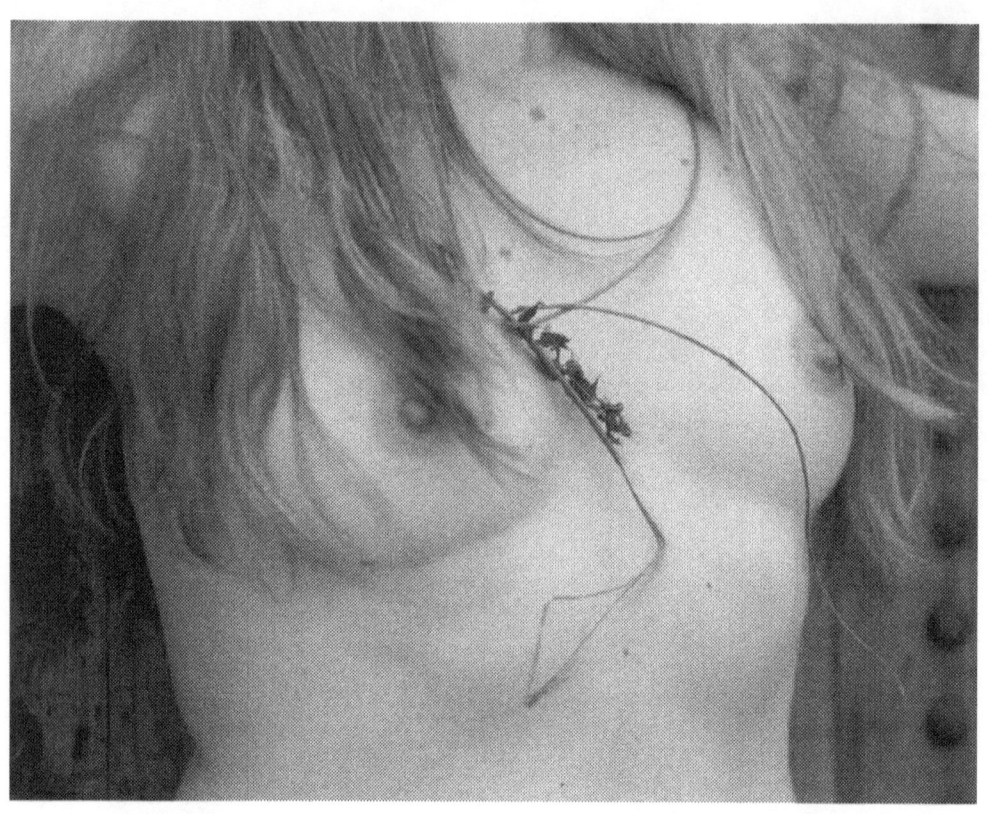

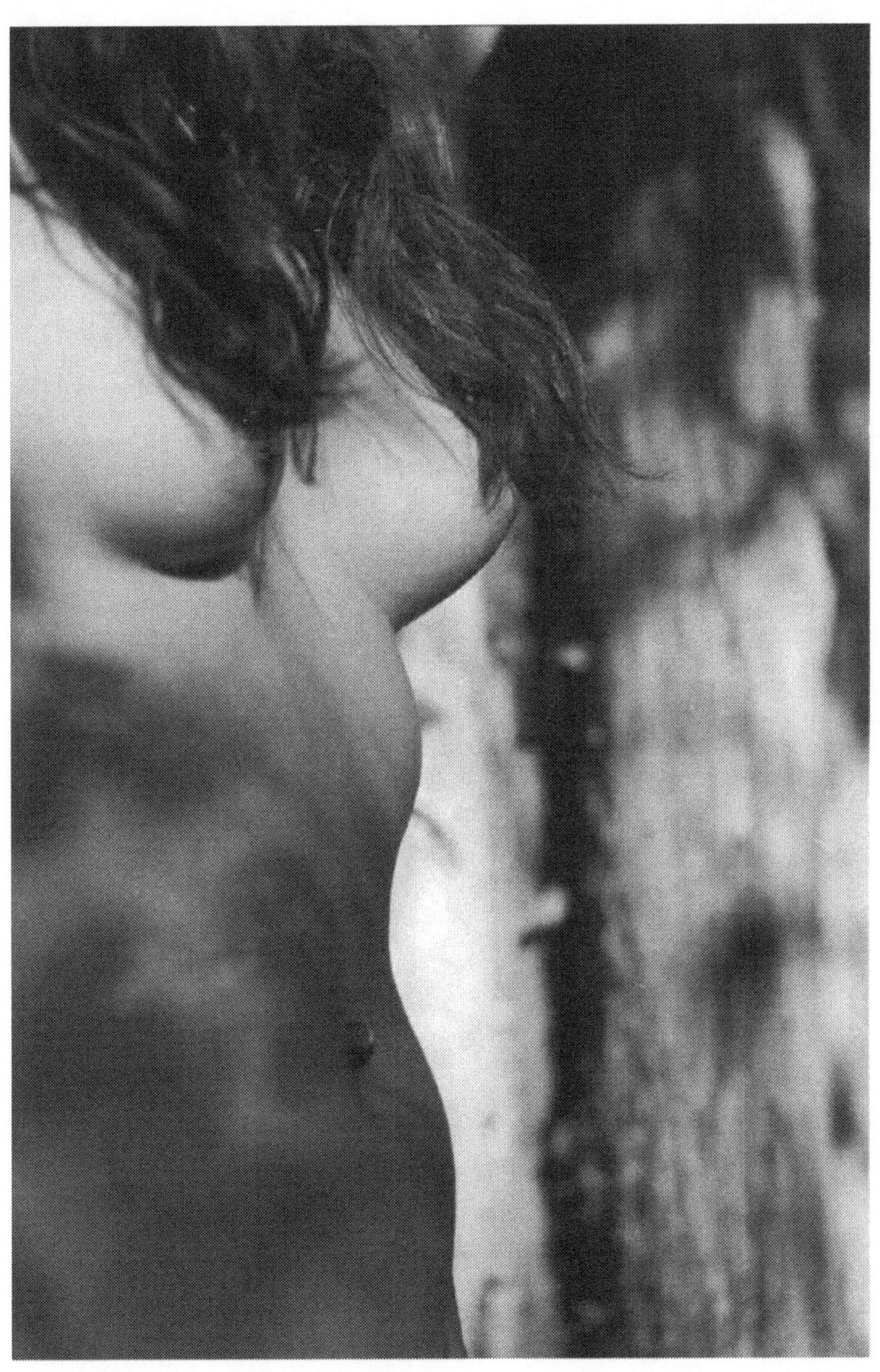

Steve Anchell

A model is an artist—a performing artist. A photographer is only able to record in the camera what a model is able to create in front of it. Photographers will benefit greatly by keeping this in mind when selecting a model for any project.

I have always preferred to work with models who are active in areas of the performing arts other than just still photography. If their only experience is in front of a still camera, they may have a limited repertoire upon which to call. On the other hand, if they are active in the theater they know how to emote and create expressions for any given situation. If they are active in dance, they are comfortable with their bodies and know how to use them to create geometric shapes, flowing forms, angles, and lines.

When selecting a model try to keep the following three points in mind:

Look: Is it appropriate for the assignment? If they can't present a certain attitude they'll never make it modeling *haute coutre*. If all they can do is glower, you don't want to spend a full day photographing them on the beach.

Attitude: Are they willing to assist you in making great images? Try to find a model with whom you can discuss your ideas, one who is willing to listen, try it your way, and then add their own creativity into the mix.

Body: Is it appropriate to the assignment? If they can't fill out a bikini, they're not going to photograph well in swim wear. If it's a male model you're selecting, he should at least have a few ripples showing above the trunks!

Above all else, in order to create truly inspired images, there needs to an artistic, energetic chemistry between you and the model.

Rose Bryan (model)

Call me a work horse but I really do not want to spend an entire photo shoot attempting to figure out what an artist likes through trial and error and then wondering if I am actually offering them material that pleases them or if they're just being polite because I'm naked. Preparation via one thoughtful and honest conversation can allow us to avoid wasting time with one awkward, unproductive shoot. Instead, I want to walk in feeling confident that I know what this person wants to see and having thought about how those wants translate into some specific poses. This way, we can focus on refining subtle details about a pose rather than starting in a vacuum.

Ideally, I walk into the shoot so close to the same mental page as the artist that we are able to collaborate deeply enough to find images that are new and delightful to that artist. Many artists seem to be waiting for lightning to strike so they can uncover that one model who can read their mind or just happens to give them exactly what they were hoping for. The hardest part of finding consistent success, regardless of the specific model you're shooting with that day, is learning how to articulate the visual images in your mind in a way that allows someone else to portray them in flesh and blood.

I will not work with any artist until we have had a meeting. Artists always assume this is so that I can make sure they're not an ax murderer but the meeting is actually the most critical component of my preparation. The best, if not only, way I know to help me understand what another human being is seeing in their head, figure out what a satisfying image looks like for them and learn what I can do to make them happy, is to review their work in their presence so that we can have a true dialogue about what they're seeing. I will start the dialogue with some broad, probing inquiries that help me narrow down an artist's genre, style, and personal taste. The first is almost always, "So what is it about these images that most excites you? Is it the lines, the shadows, the colors, the emotional affect or is it something more narrative?" I am still shocked when an artist tells me that I am the first model who has asked them this question.

Jeff Dunas

Every student has the same question, "How do you know how to pose the model?"

The answer is insanely simple! *Say Something.* I notice at my workshops that the student photographers shoot at high speed, blasting away without ever saying anything to the model/subject, while the model is just sitting there, posing as they see fit. Interaction is the key. Let the model strike a pose if you like, but then begin to suggest adjustments and before you know it you're saying, "Great–HOLD IT!" As people move they pose in ways that seem comfortable to them, but these poses don't always look comfortable. One must make suggestions that render the pose comfortable looking. Another thing I tell students often, above all else, in fact, is remember, you're transposing something three dimensional into something two dimensional. That hand creeping up onto the stomach might not even look like it's the model's own hand in the picture! That leg behind the other might look normal to the eye, as it's behind the other. In the picture, however, the leg closest to the camera simply looks misshapen because of the leg behind. Look for body parts–extremities–and find natural positions for them. You must be able to sign your name to every centimeter of the frame. So, I recommend a tripod at first, even most of the time, because it's the only way to really take one's time, examine the frame, try to see it as a two-dimensional picture through the viewfinder.

Quite simply, I look for models who appeal to me. In some form or another, the model should appeal to the artist. The model should be visually appealing to the photographer. This doesn't mean pretty, only appealing in some form. That's the starting point for the picture.

There is no perfect model. There is no perfect photographer and there is no perfect picture. It's what you make it. That was true before Photoshop and it is still true in a Photoshop world. It's in *your* head.

Unique Poses

Sometimes the most difficult poses to attain are those that look the simplest. While some of the following poses may not look unique, they are the ones that may have been caught when the model least expected it. Perhaps it was a slow turn of the body, or a time of rest; these are the times when you have to be at your sharpest. These are also the times when you may create the most compelling images.

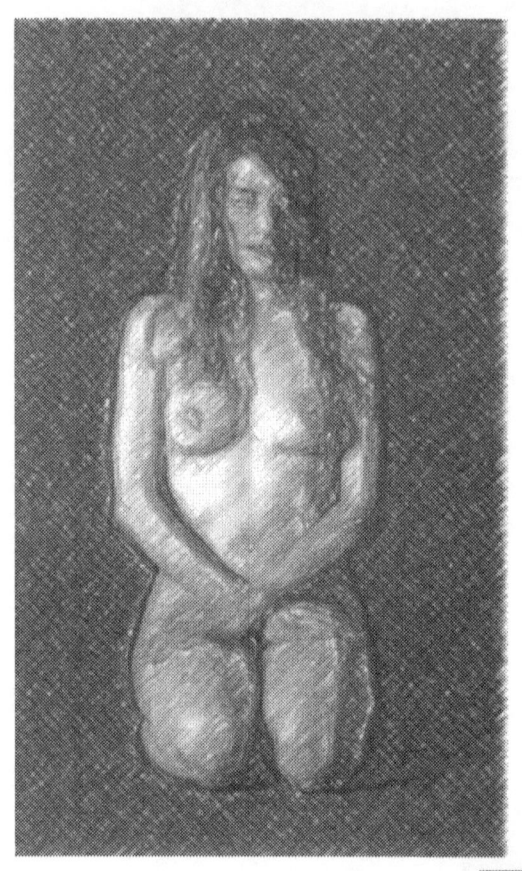

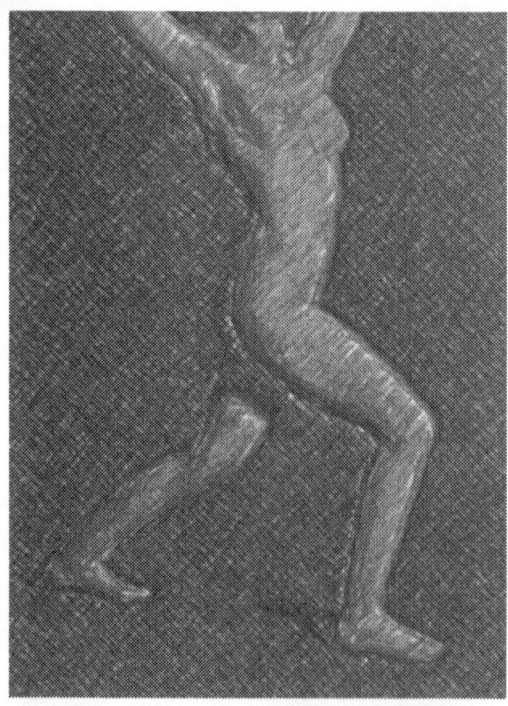

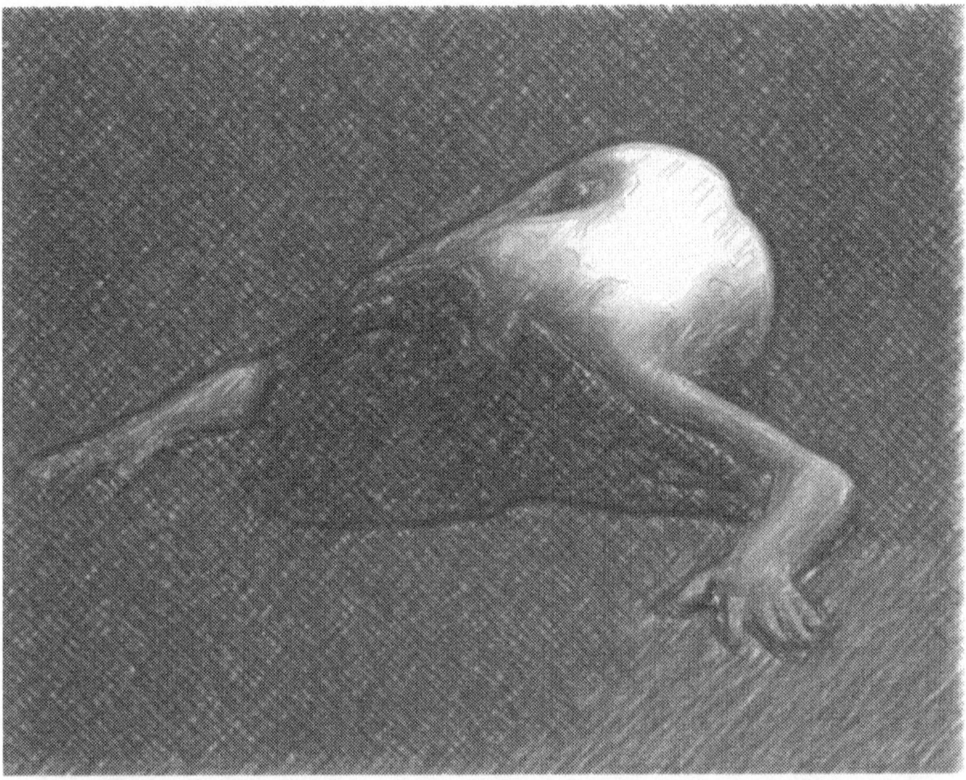

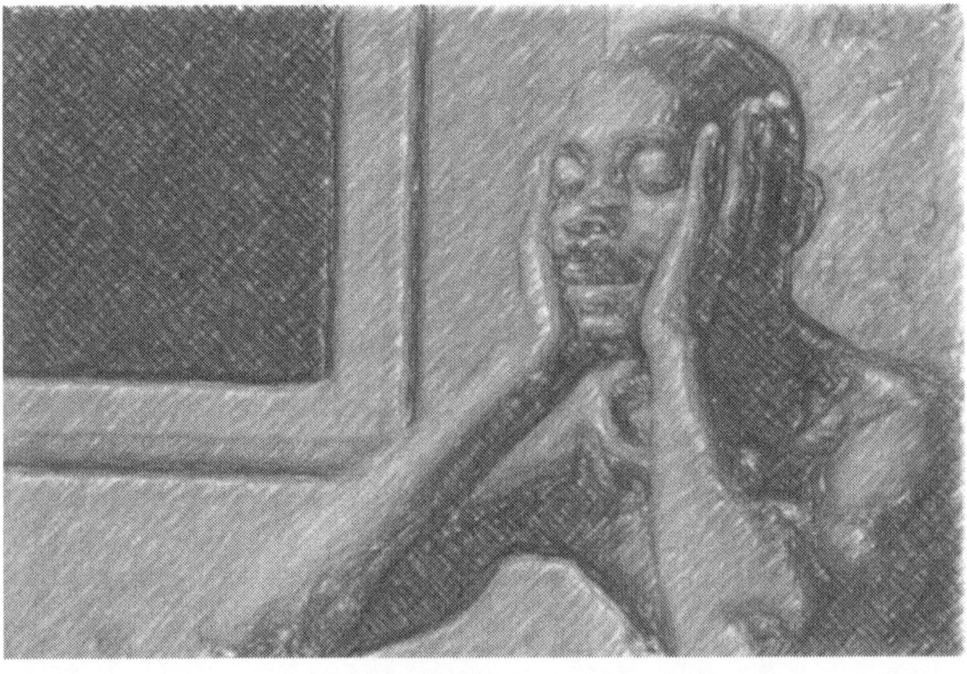

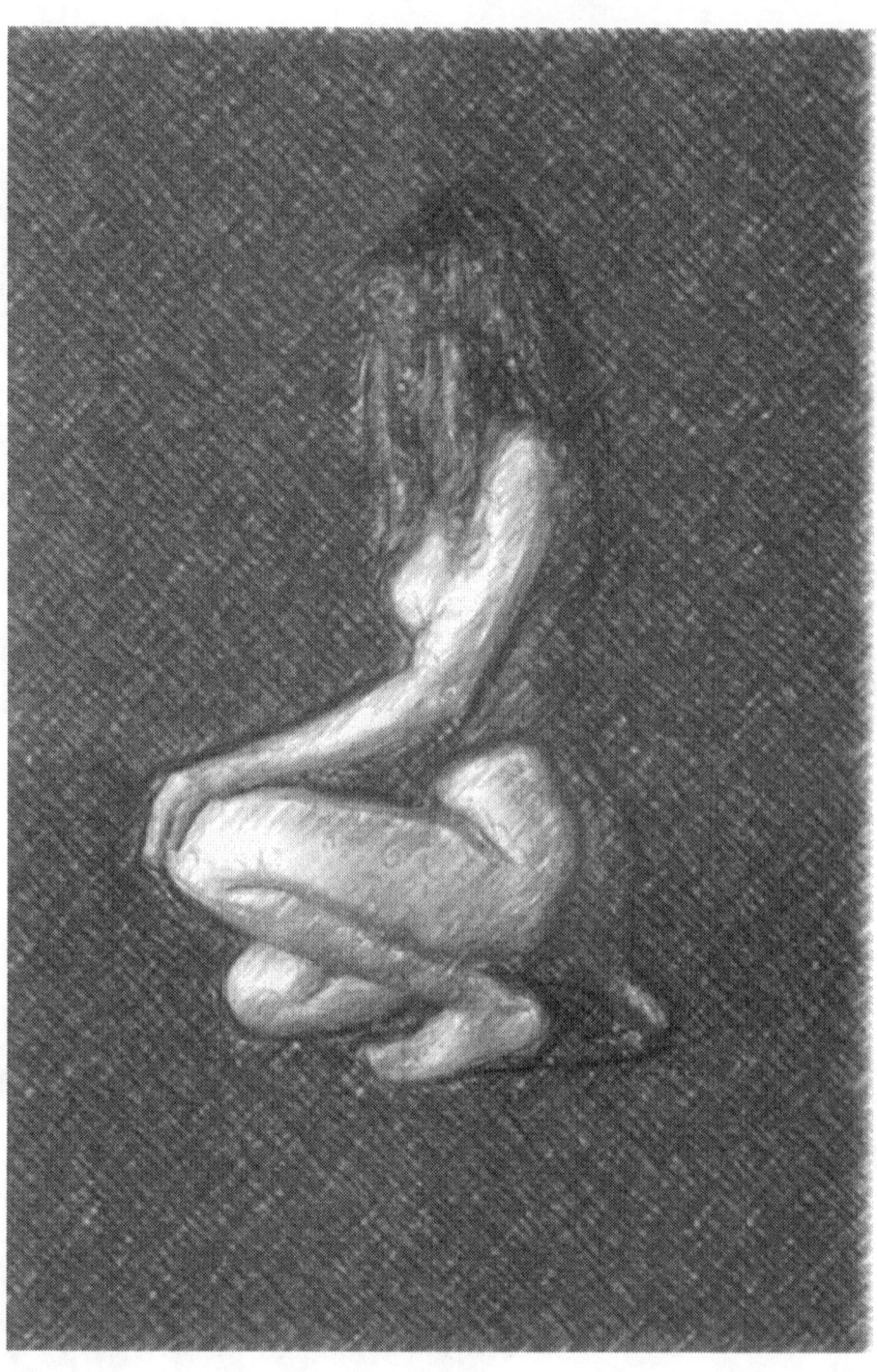

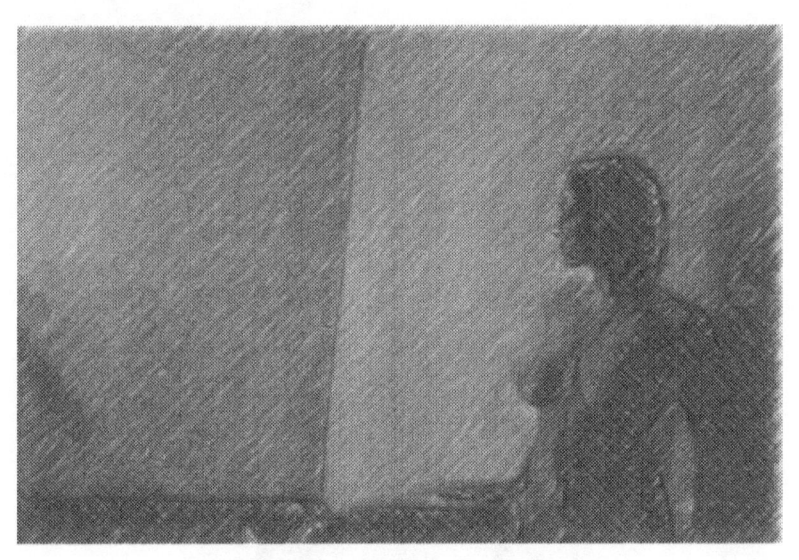

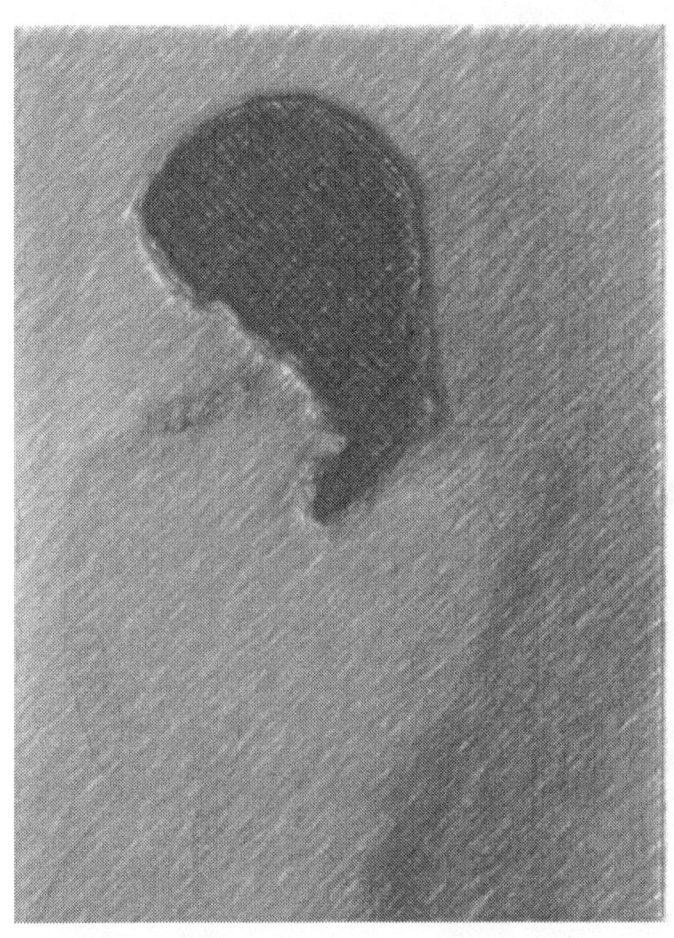

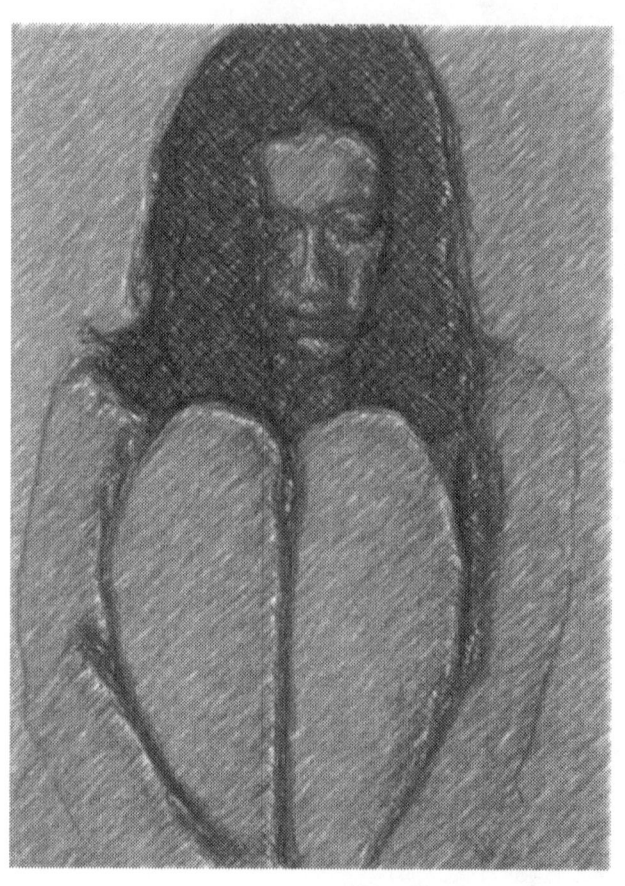

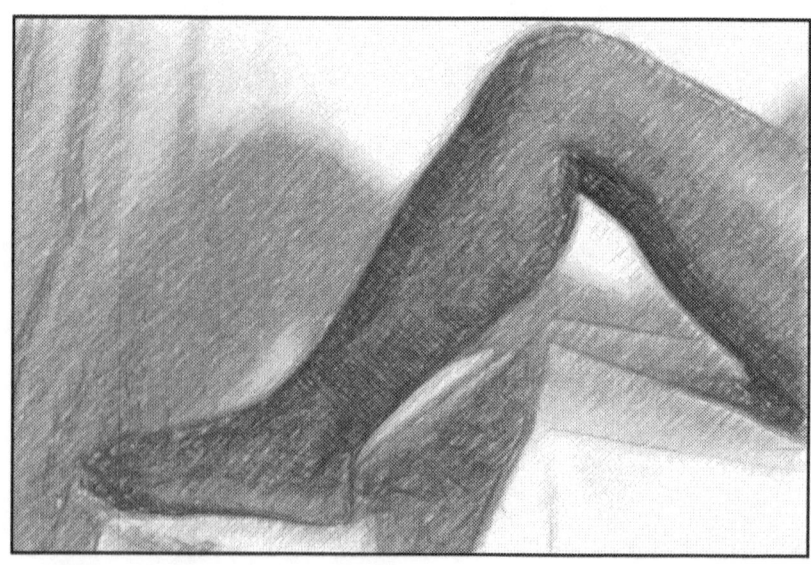

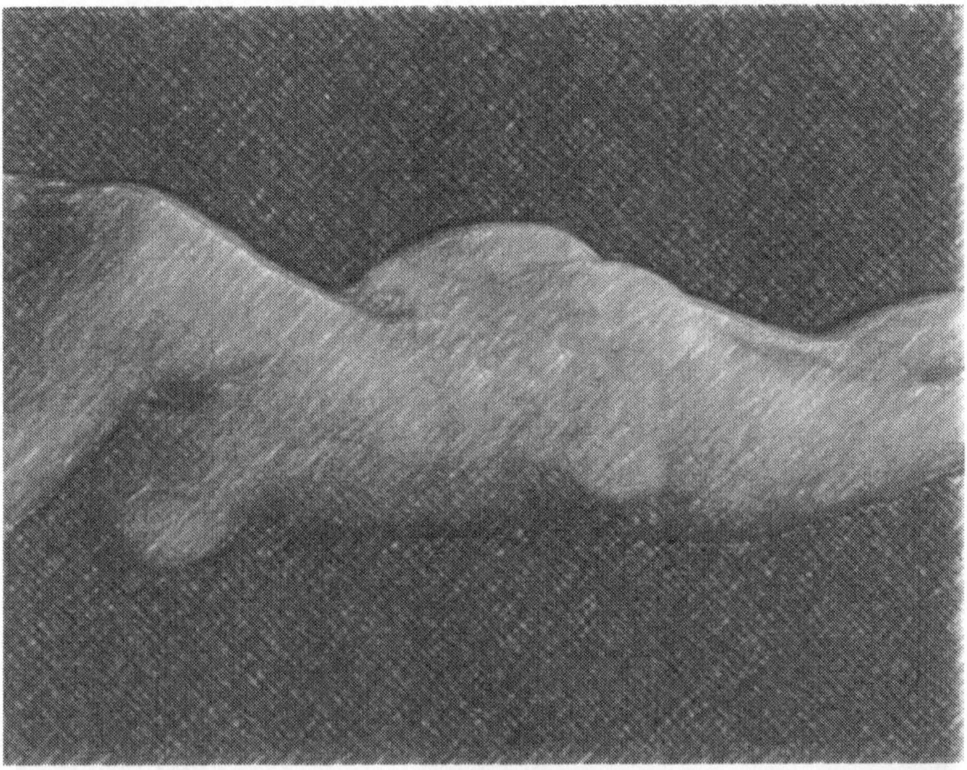

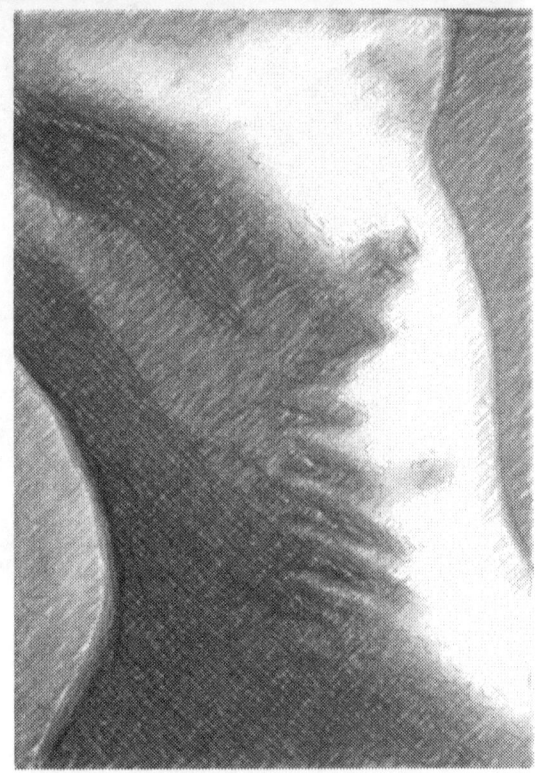

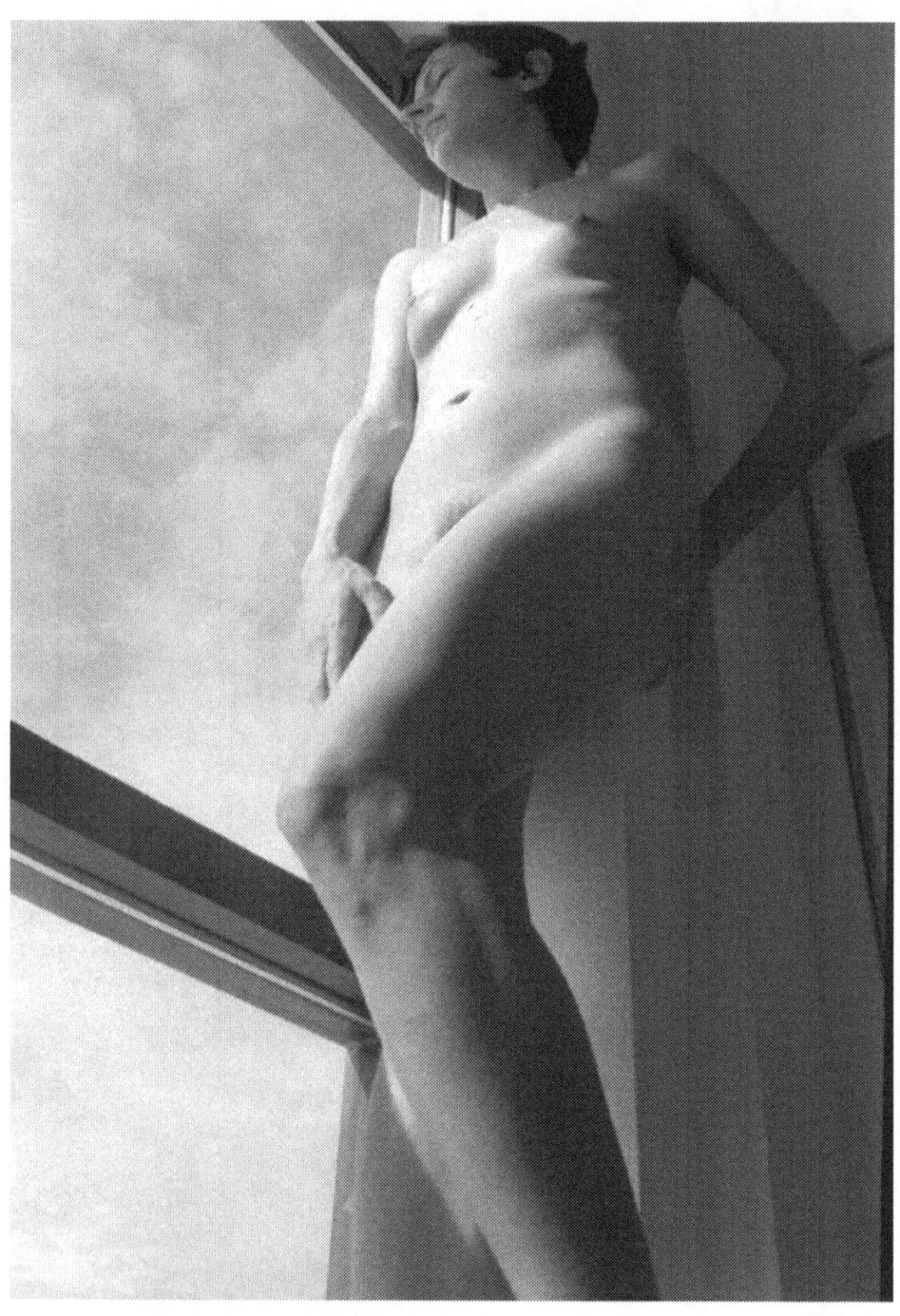

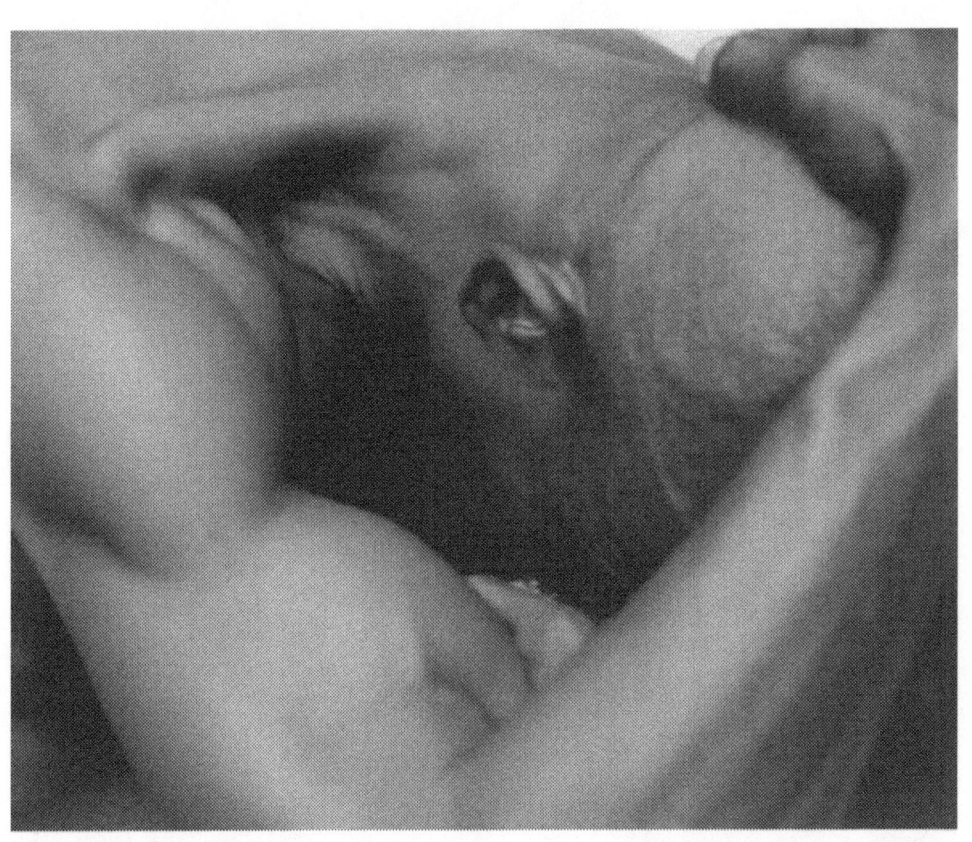

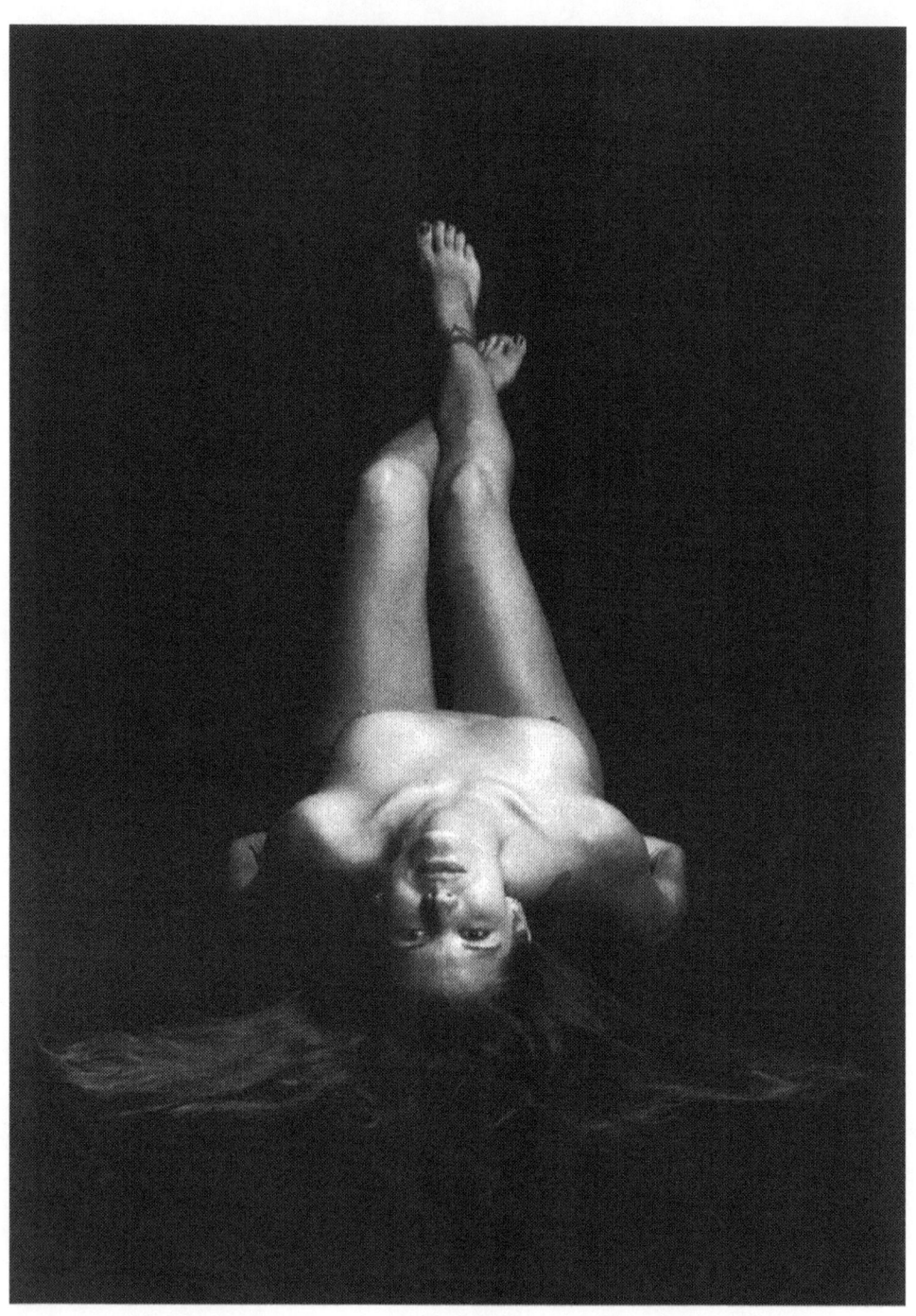

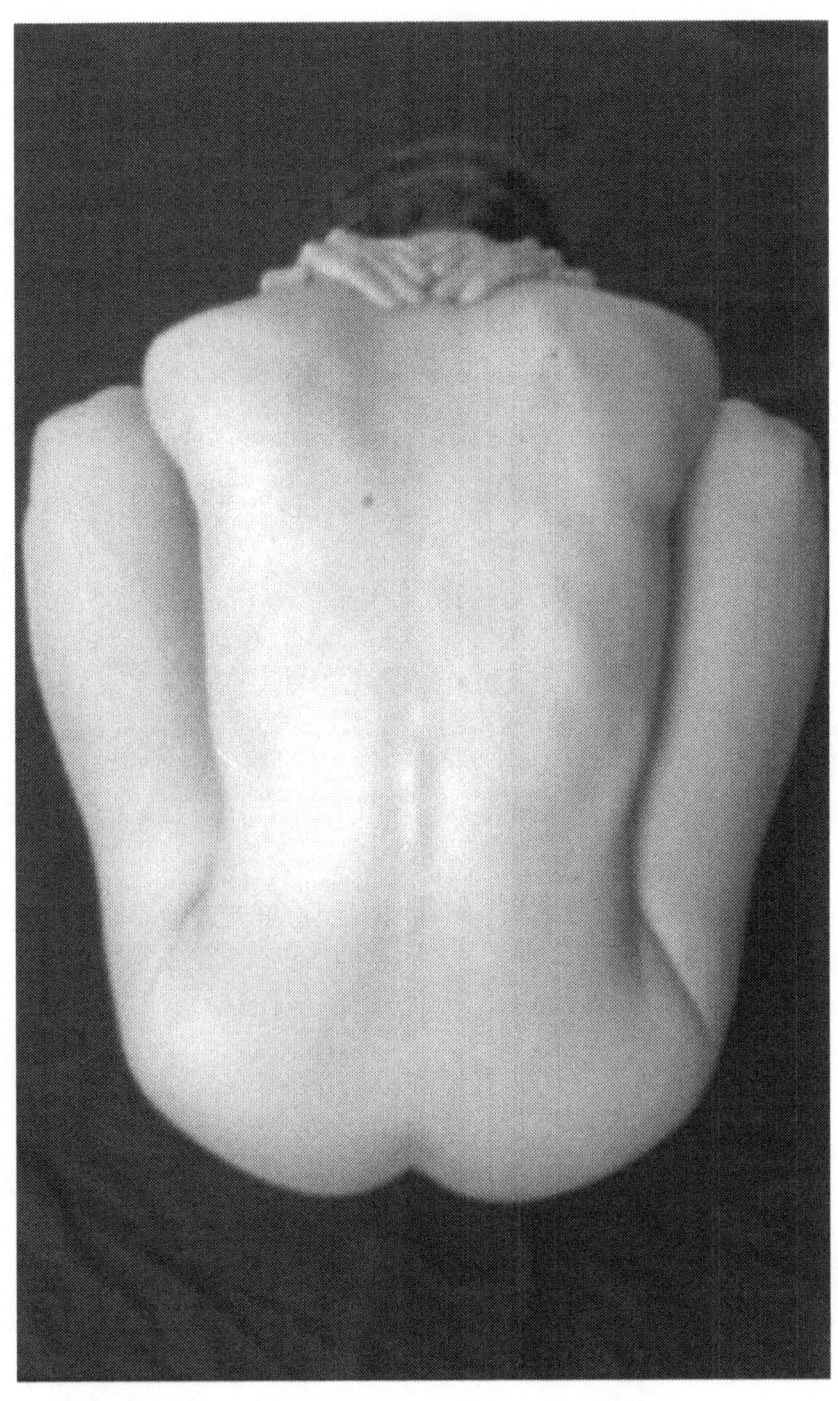

Robert Farber

When working with a model for nudes, whether I've worked with them before or not, I have them come on to the set usually in a robe or other loose clothing. I want to make sure any elastic or clothes marks are not present, but mostly so I can progress into having them slowly become nude. I first shoot some head shots, to make them feel good about what I am doing and then have them start baring their shoulders until I progress to the point of nudity that I am looking for.

I move around the nude model usually silently, while gently directing them, if necessary, letting them get into the mood that I am looking for. This is usually handled as if it is their private moment. I try not to invade their private moment, as if I was not there. I usually do not show the face, so that the viewer does not relate my subject to a particular person, and there is more of a mystique.

Elizabeth Opalenik

Throughout my career it has always looked as though 36 people held the camera for a roll of film. Everything interests me, as does every model.

I am interested in the beauty within that each of us carries. I think as an artist, it is your viewpoint on that beauty that reveals your creativity, not the package that the model arrives in.

We are our perceptions. In how they are made and how they are viewed, photographs (drawings, sculpture, etc.) are self-portraits. They are a journey into the possibility of what else there is. I have been known to ask people on the street, in restaurants or in my classes, to model for me. Let the images that you have made previously speak for you. If you have integrity in your work, you will have models.

Abstract

To shoot a model in an abstract style, is to not photograph the person as a whole, focusing, rather, on the elements of the body that form geometric crossroads, places where line and form, shadow and light meet.

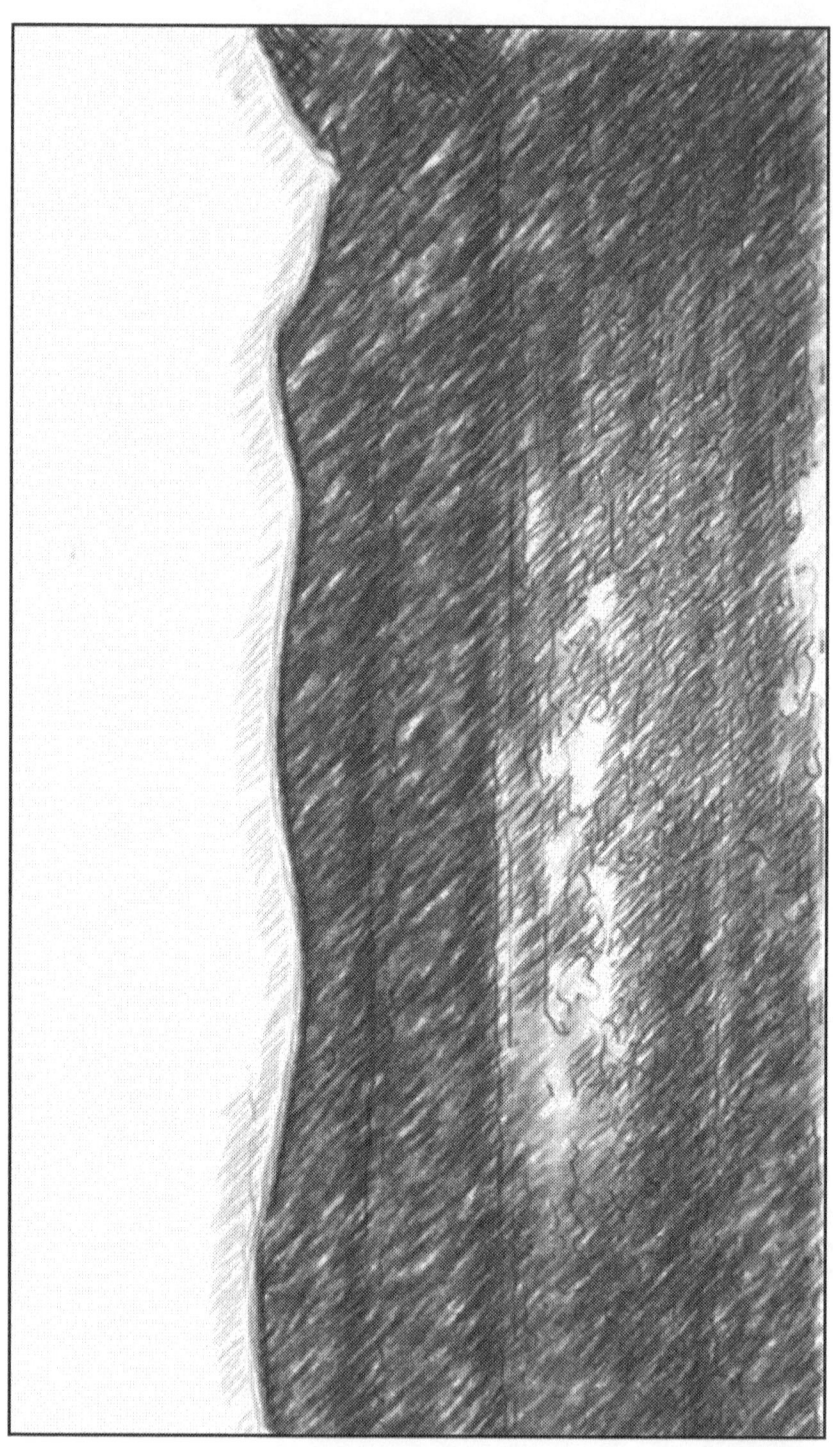

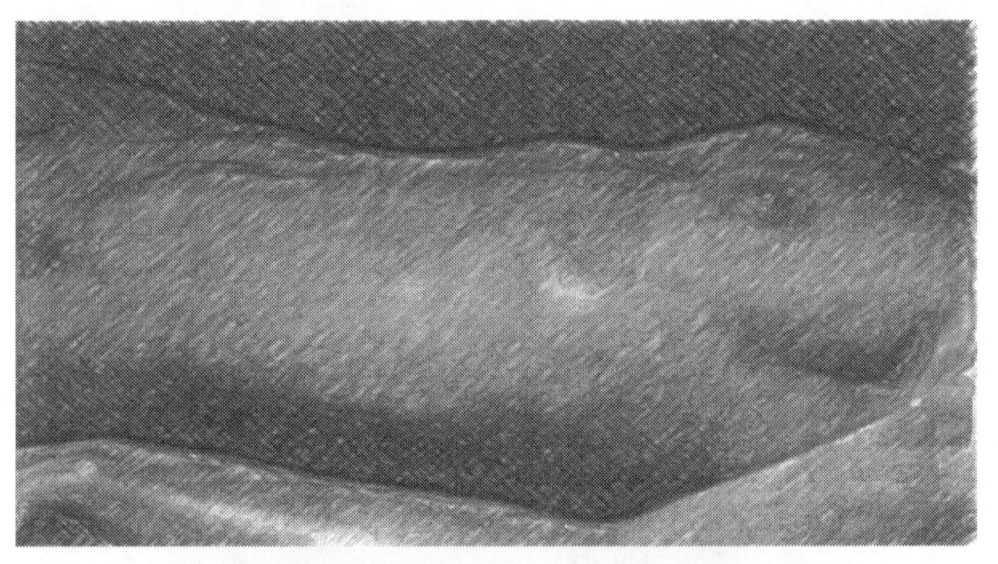

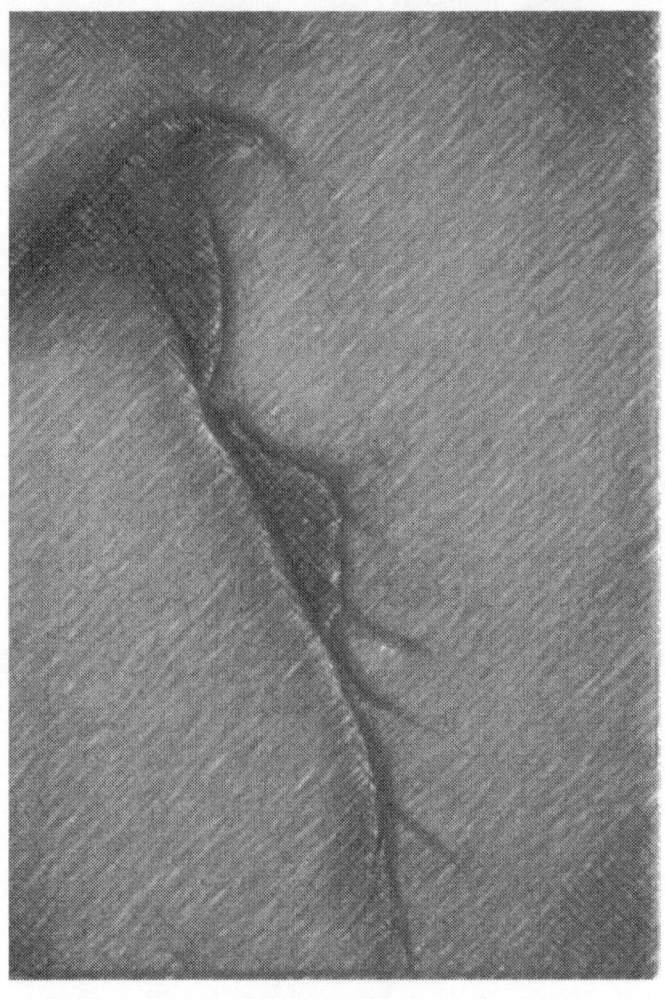

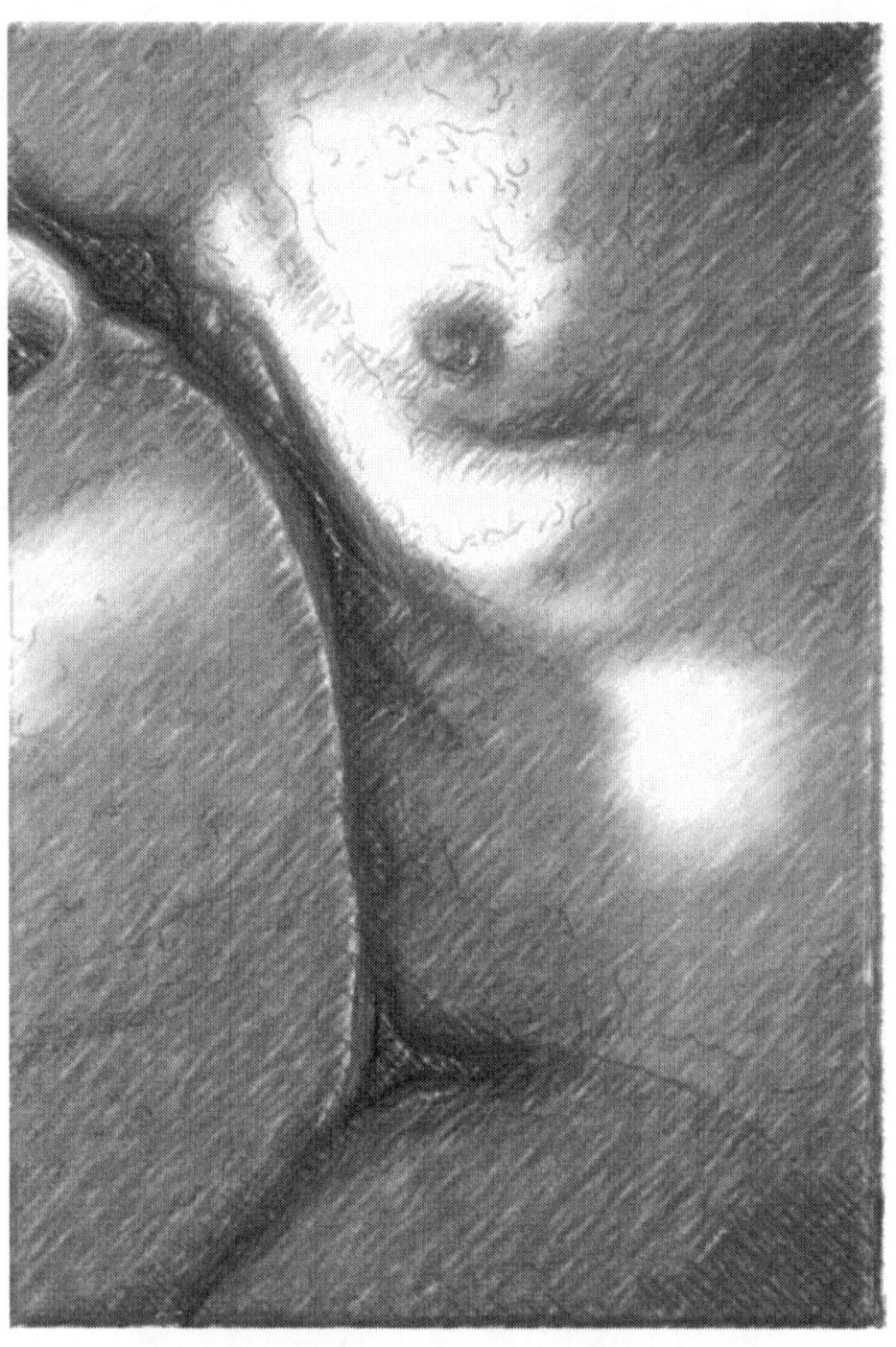

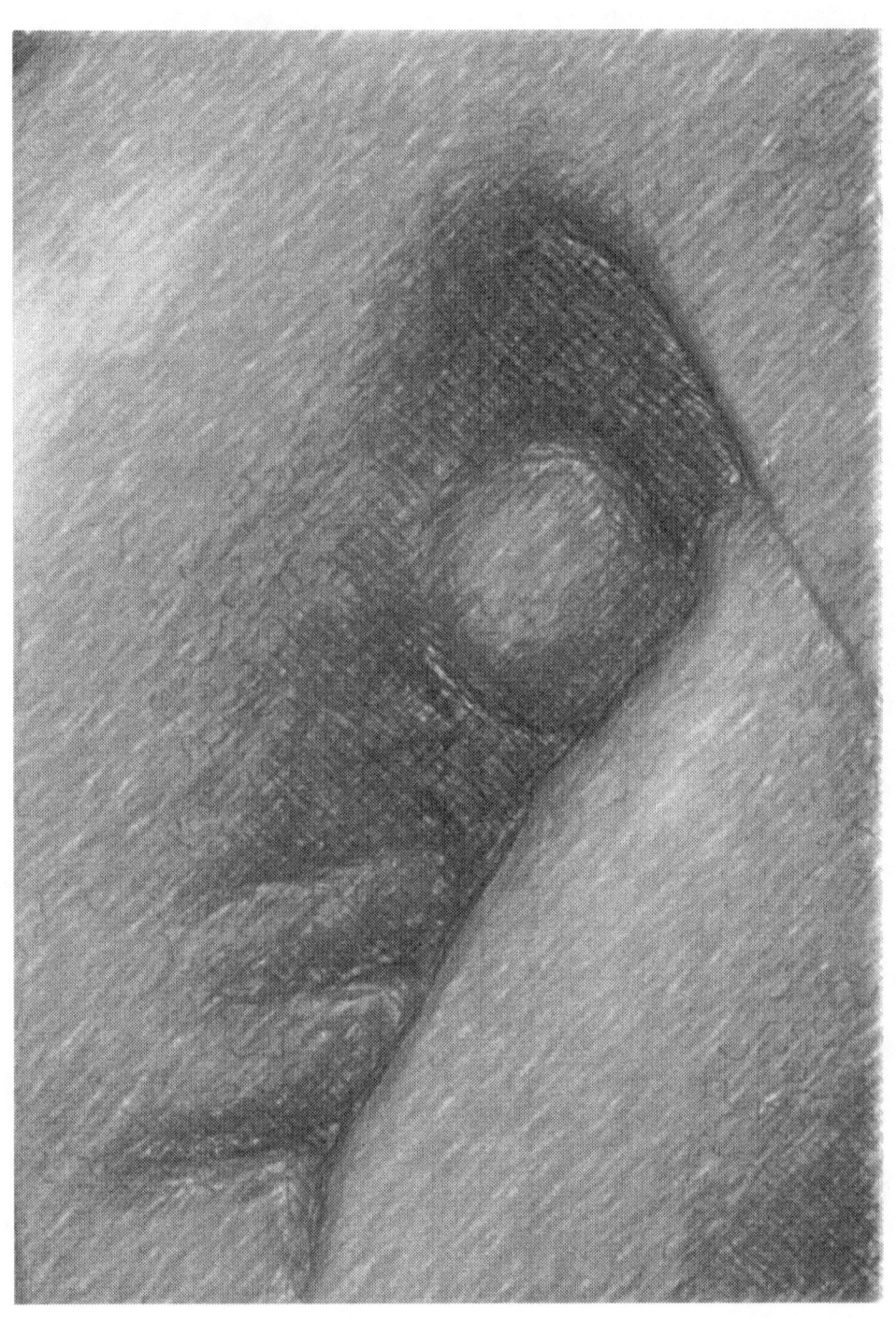

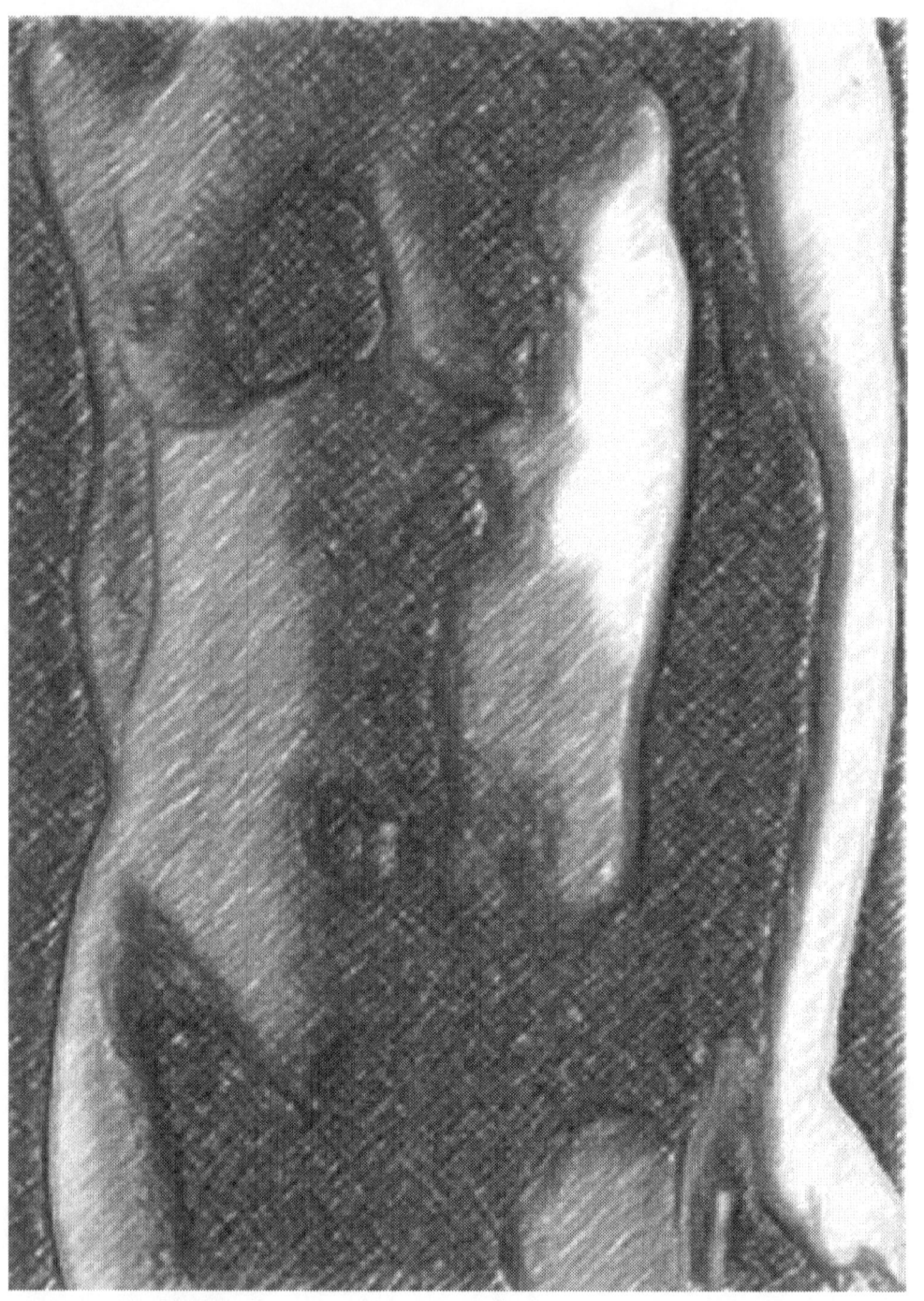

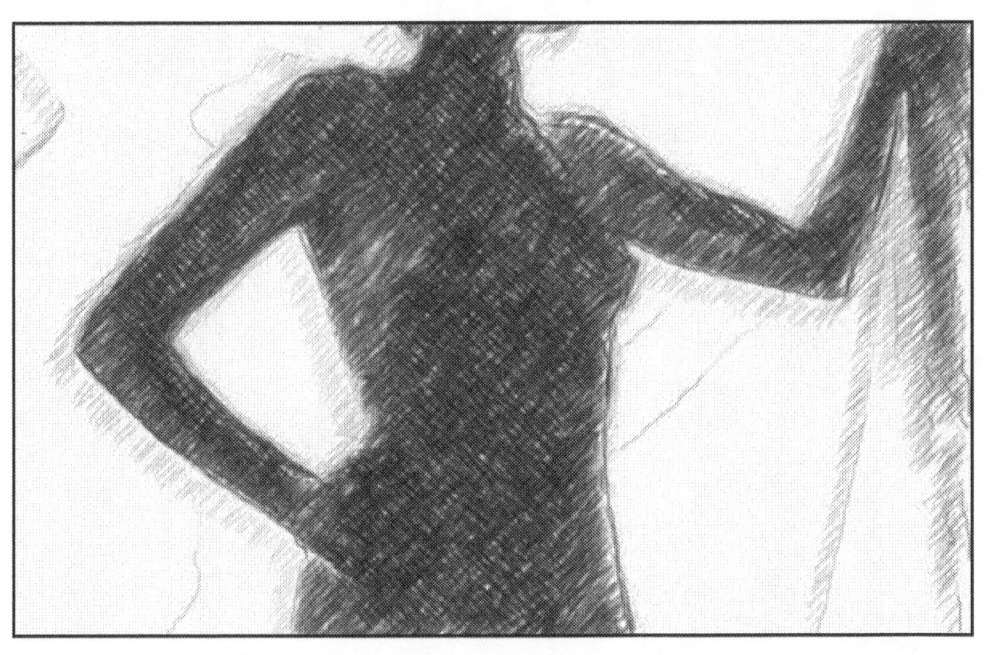

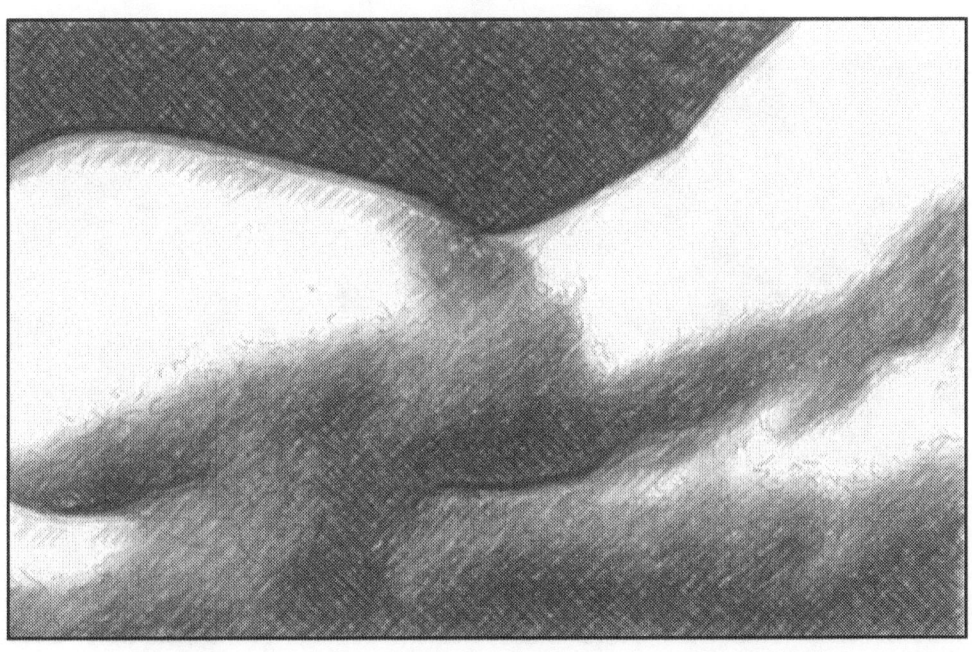

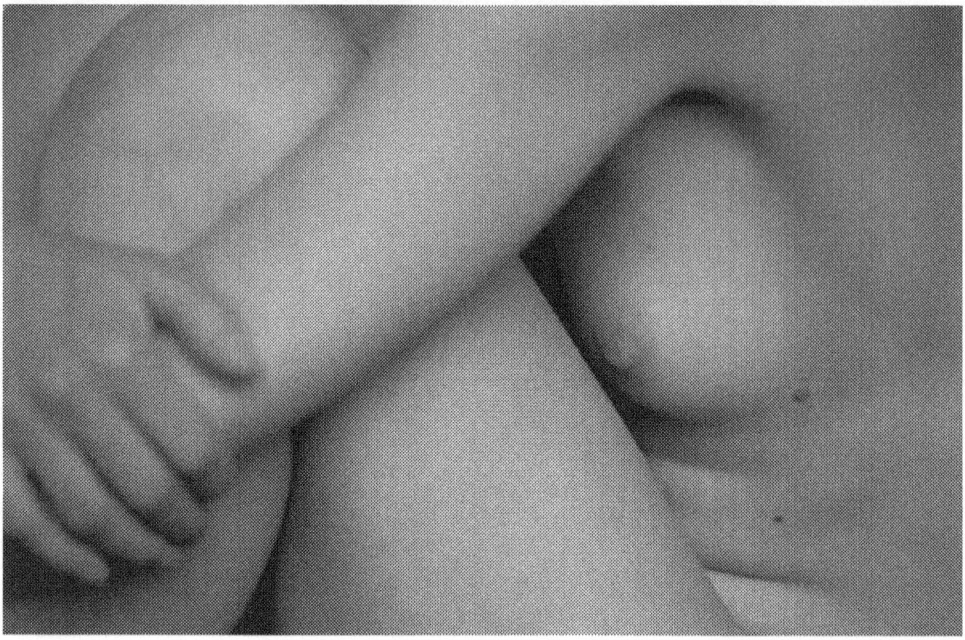

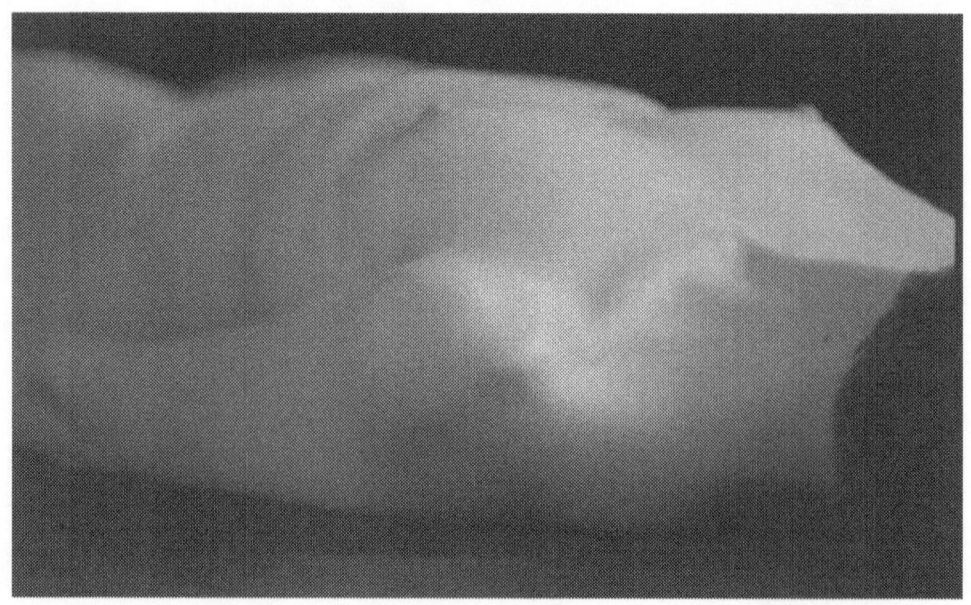

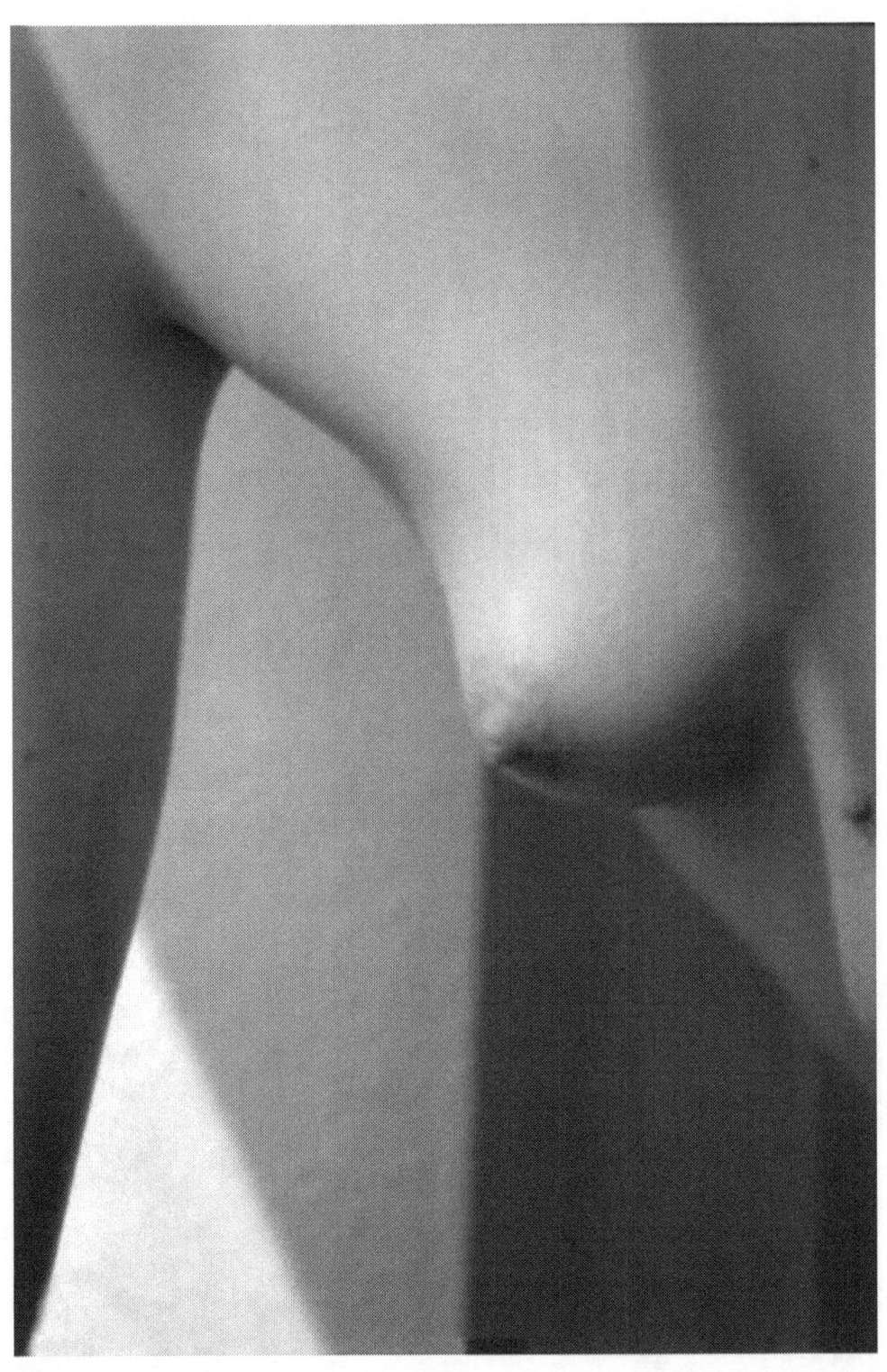

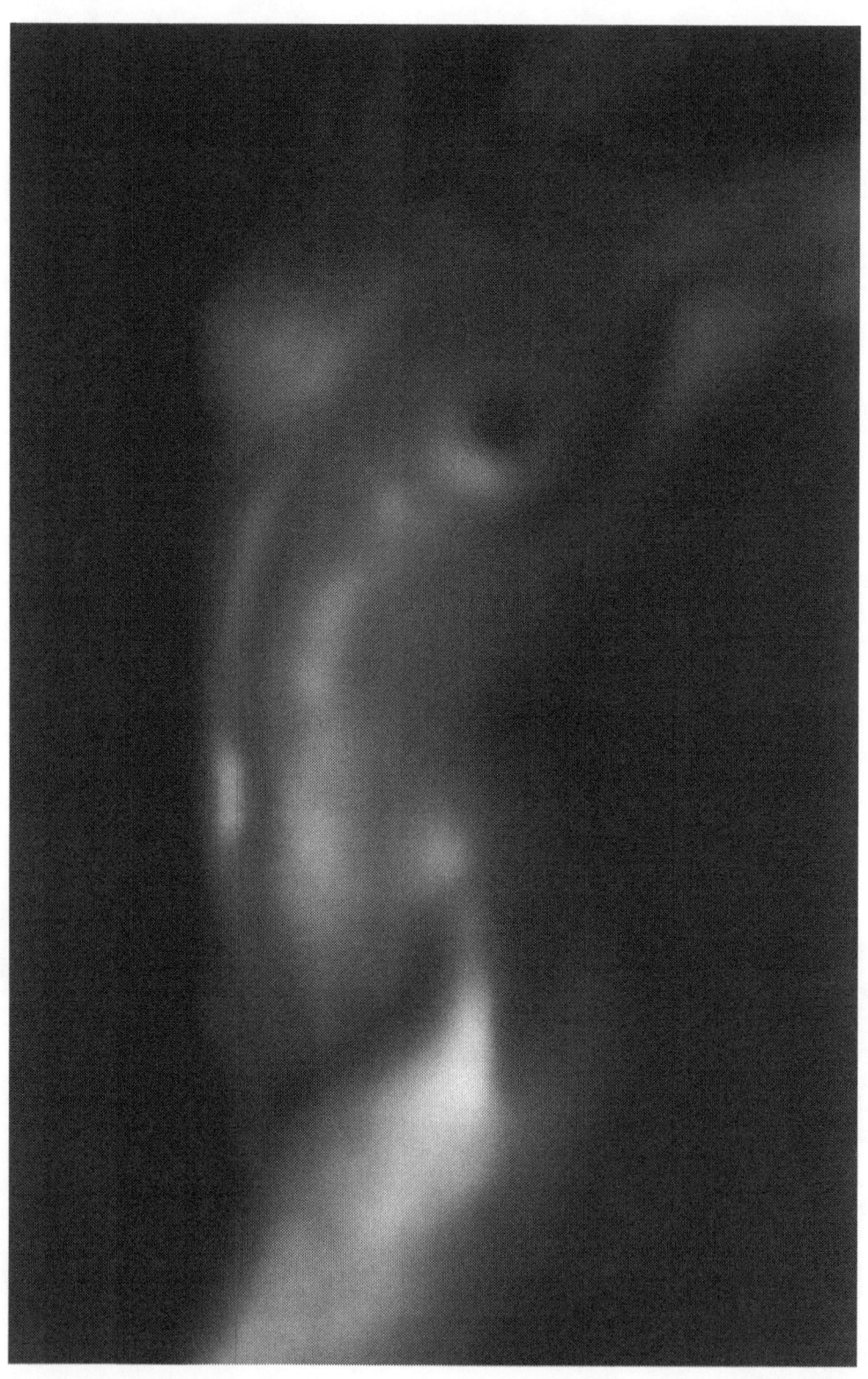

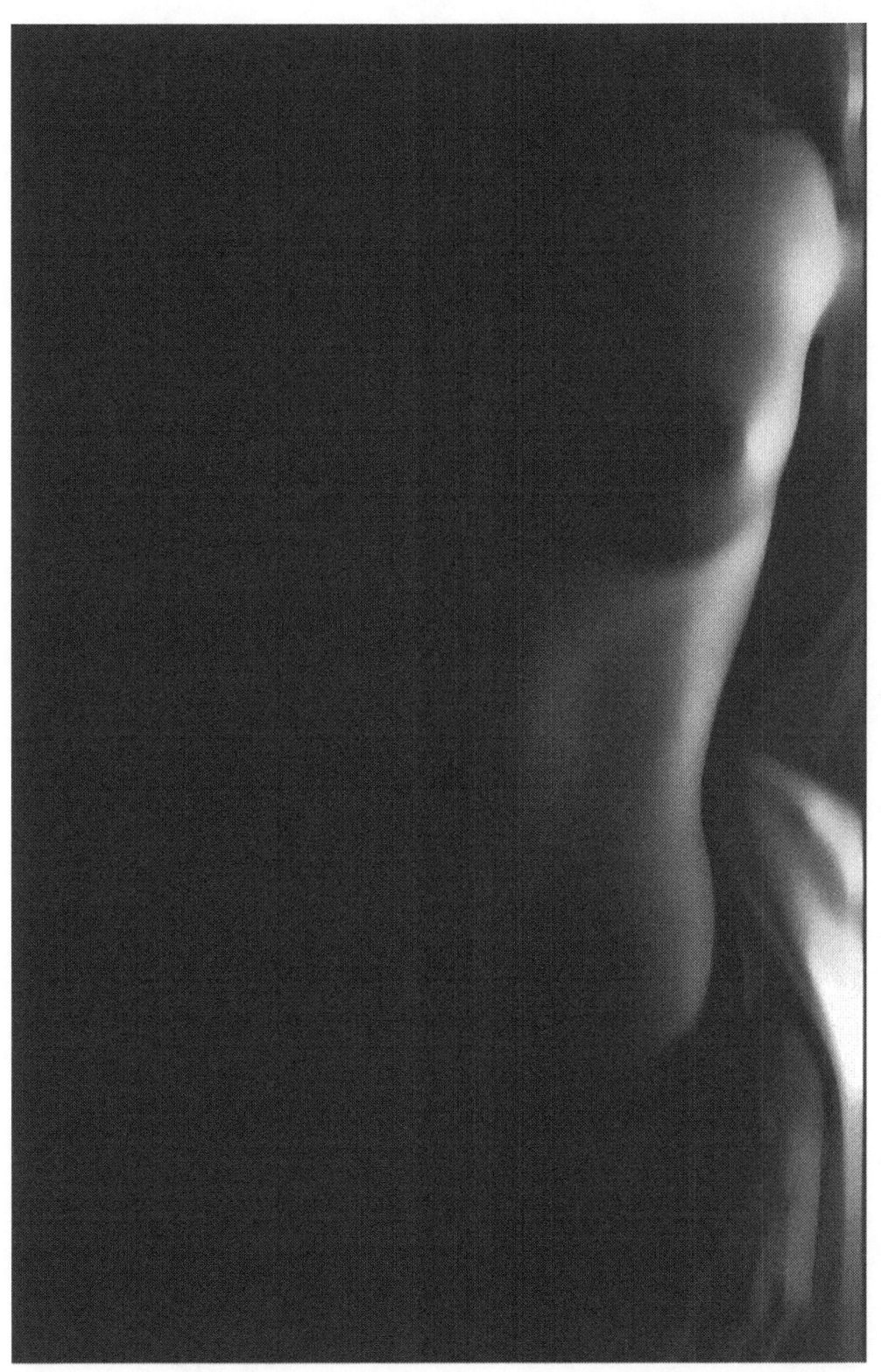

Ashley Rae *(model)*

One of the key elements for a successful shoot is good communication. As a model, I like to know what types of images the photographer wants to shoot. When I know what is expected from me, I can be better prepared for the shoot, both physically and mentally. For example: bringing appropriate clothing, having proper makeup and hair styling tools, etc. It is very helpful to meet beforehand and go over ideas, especially if we have just met! I find that the best photo shoots are with people I have worked with in the past. The better we know each other, the more productive the photo shoot. With good communication, everyone can have realistic expectations and better preparation.

Bill Lemon

Posing the female form is like driving a well-balanced sports car on a curvy road. You take every curve with caution. The female form is like a piece of artwork and it should be photographed in that manner.

I think that when I first started my photographic career I wanted to pose my female models to tell a story in the ambience of the scene and after 27 years I still feel the same. I look at it just a little different now, however. I've been told that the models look passive and I think that I'm shooting that feeling. I'm trying to capture the model in her elements as if I were not there.

I photograph at all times of the day to meet everyone's schedule using the light to my advantage with artistic nudes. Preferably, I like the late afternoon to sunset light for the warm tones and a dramatic look in both my casual as well as my artistic nudes.

Kate Taylor *(model)*

My background in dance, acting, and choreography has enabled me to utilize those skills in my modeling pursuits. I think they allow me to bring so much more to each session, including offering the photographer a wider selection of poses and styles to work with. I am always looking to grow in my career, and traveling and working with so many talented people has enabled me to do just that.

I consider my work to be like a dance piece...a creation and performance to the camera. Like my choreography, I like to keep my ideas fresh. I go with the attitude that today's work will soon be yesterday's work. Therefore, I'm always looking for a new direction in which to take my work, and the variety of photographers and artists I have worked with helps me continue in that process.

Accessories

On many occasions, especially when you are using a model for the first time, it may be difficult for her to pose without any clothes on. This is a natural reaction for a novice model. Frequently, all you need to do to relax the model is to let her wear a few articles of clothing or jewelry when the session first begins. Accessories can also add interest or intrigue to an otherwise common pose.

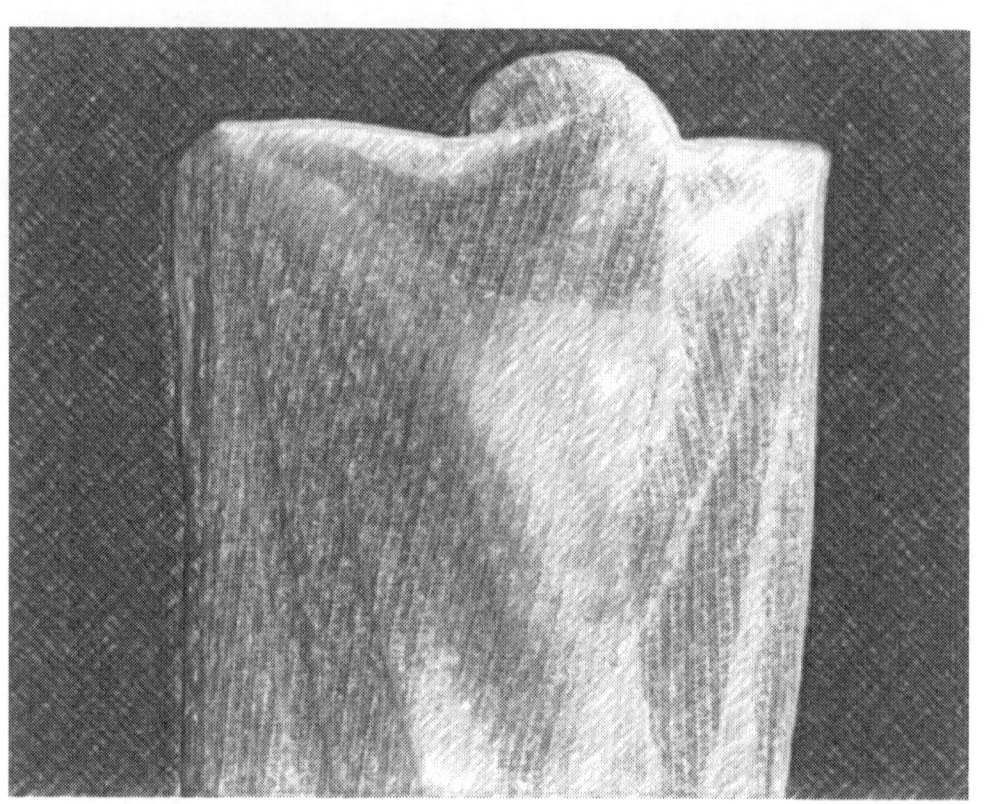

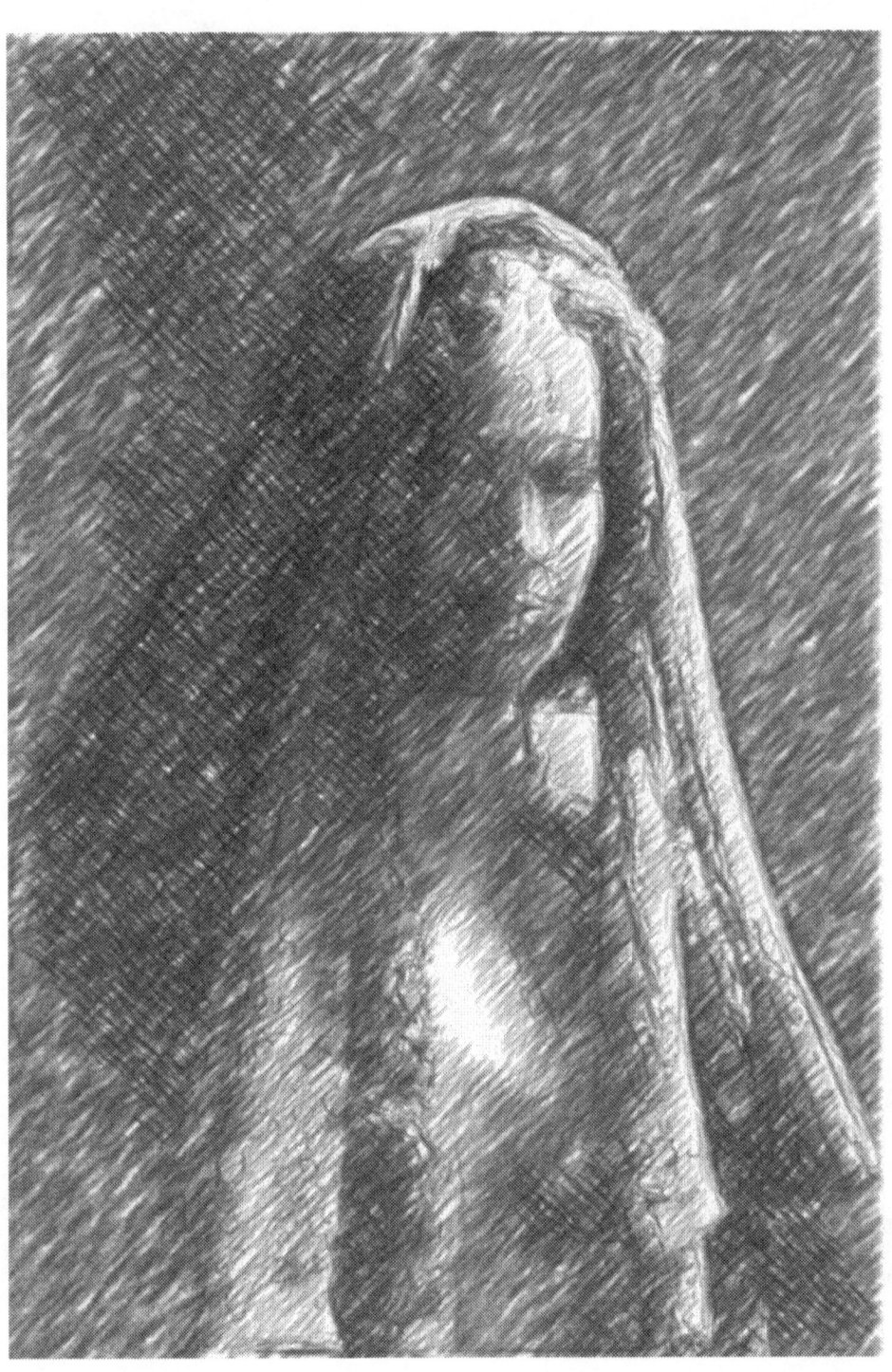

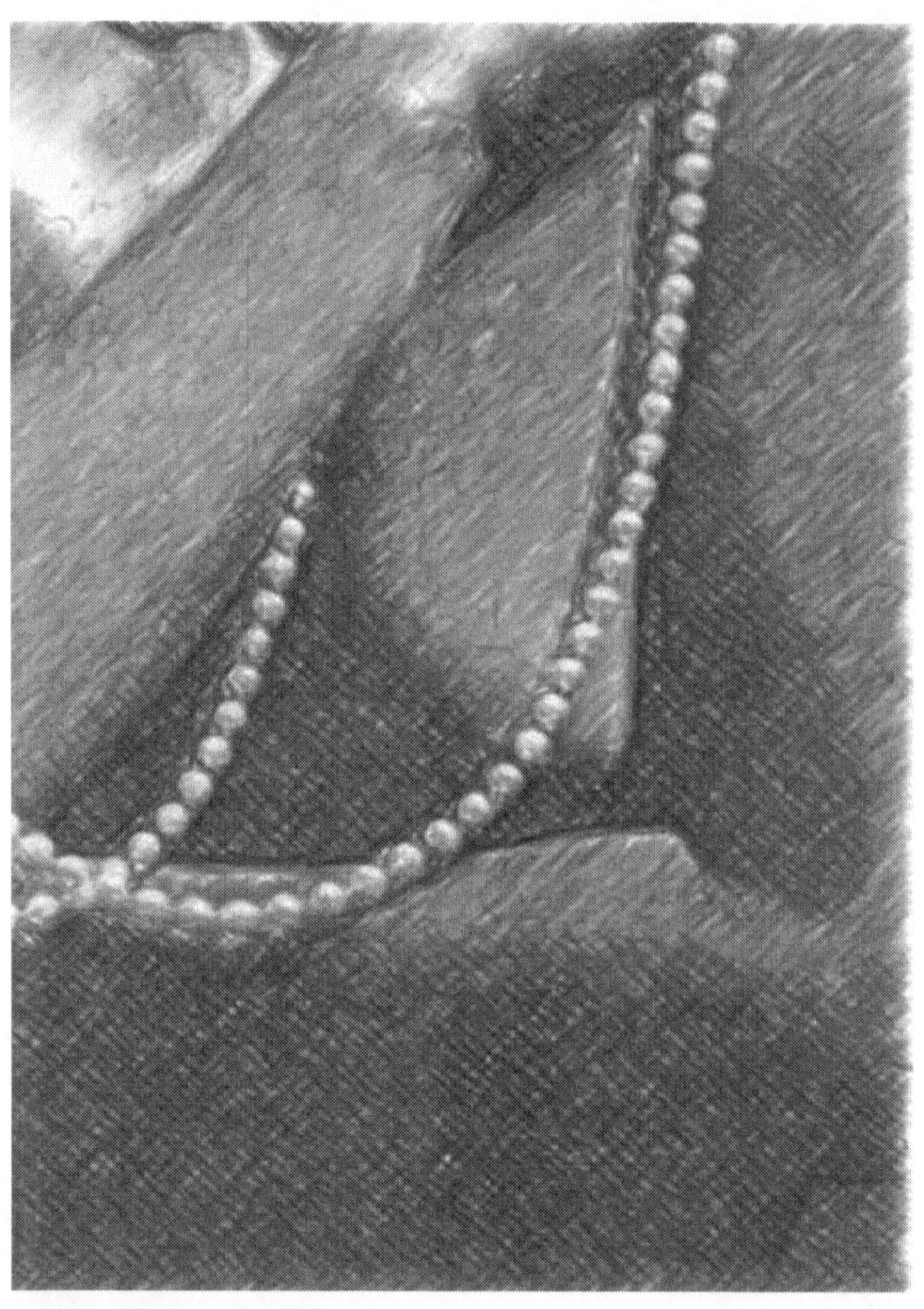

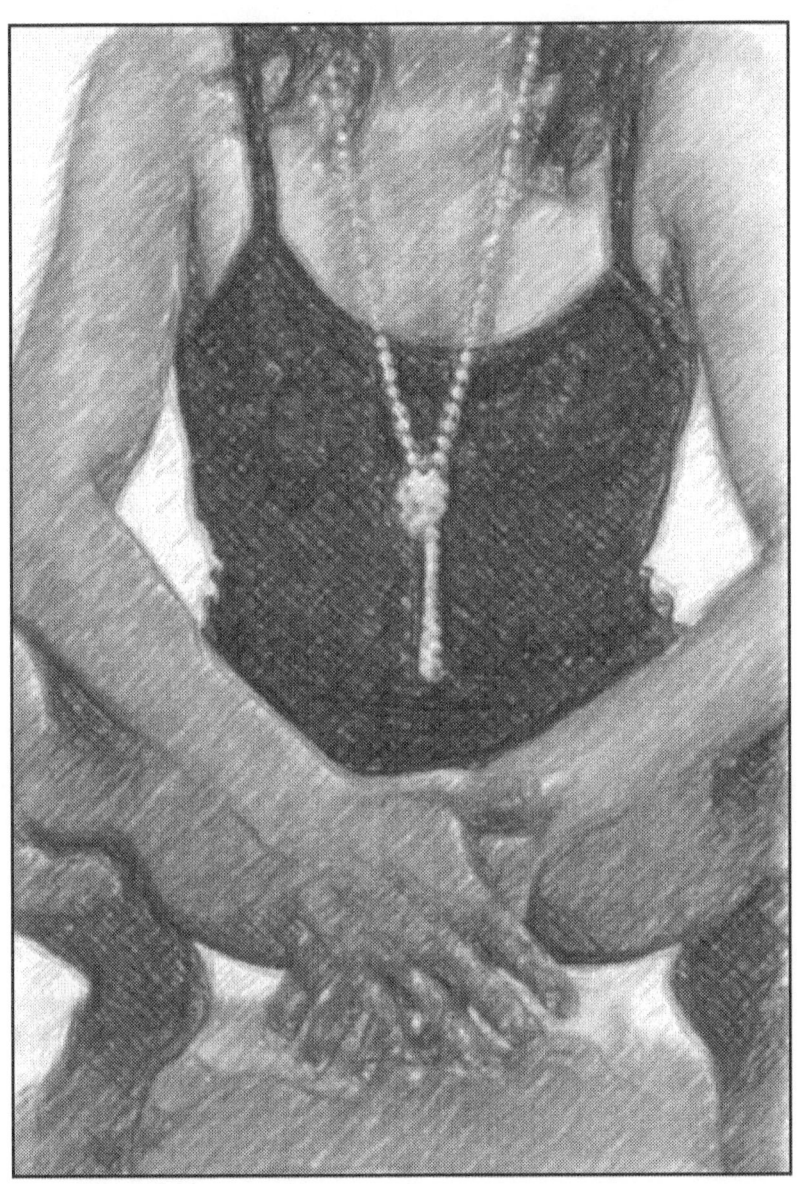

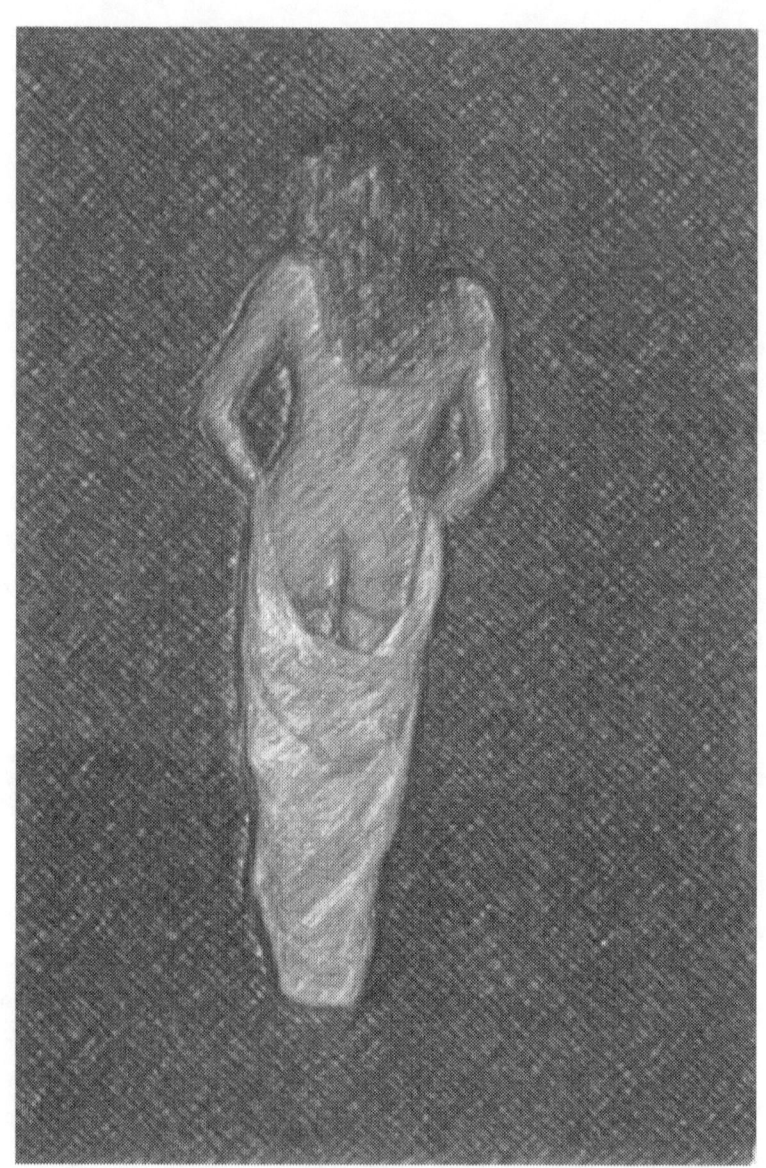

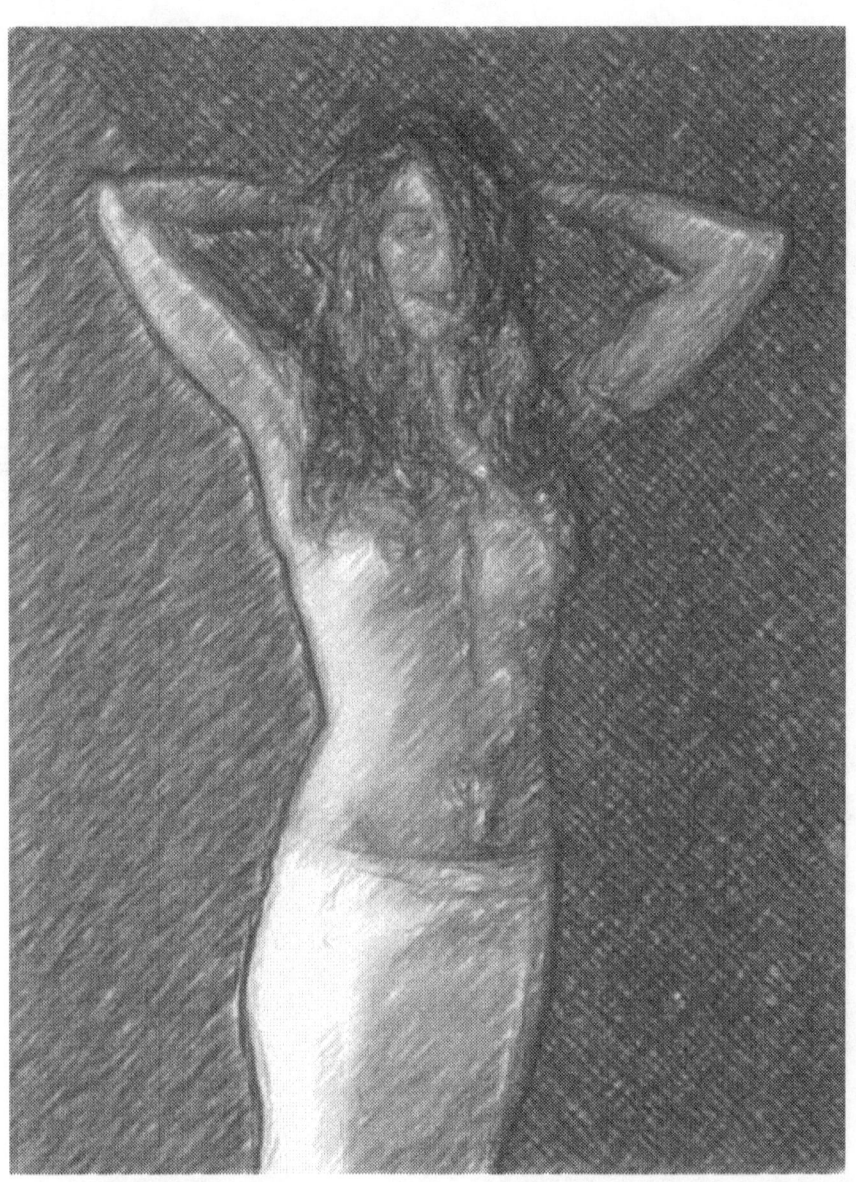

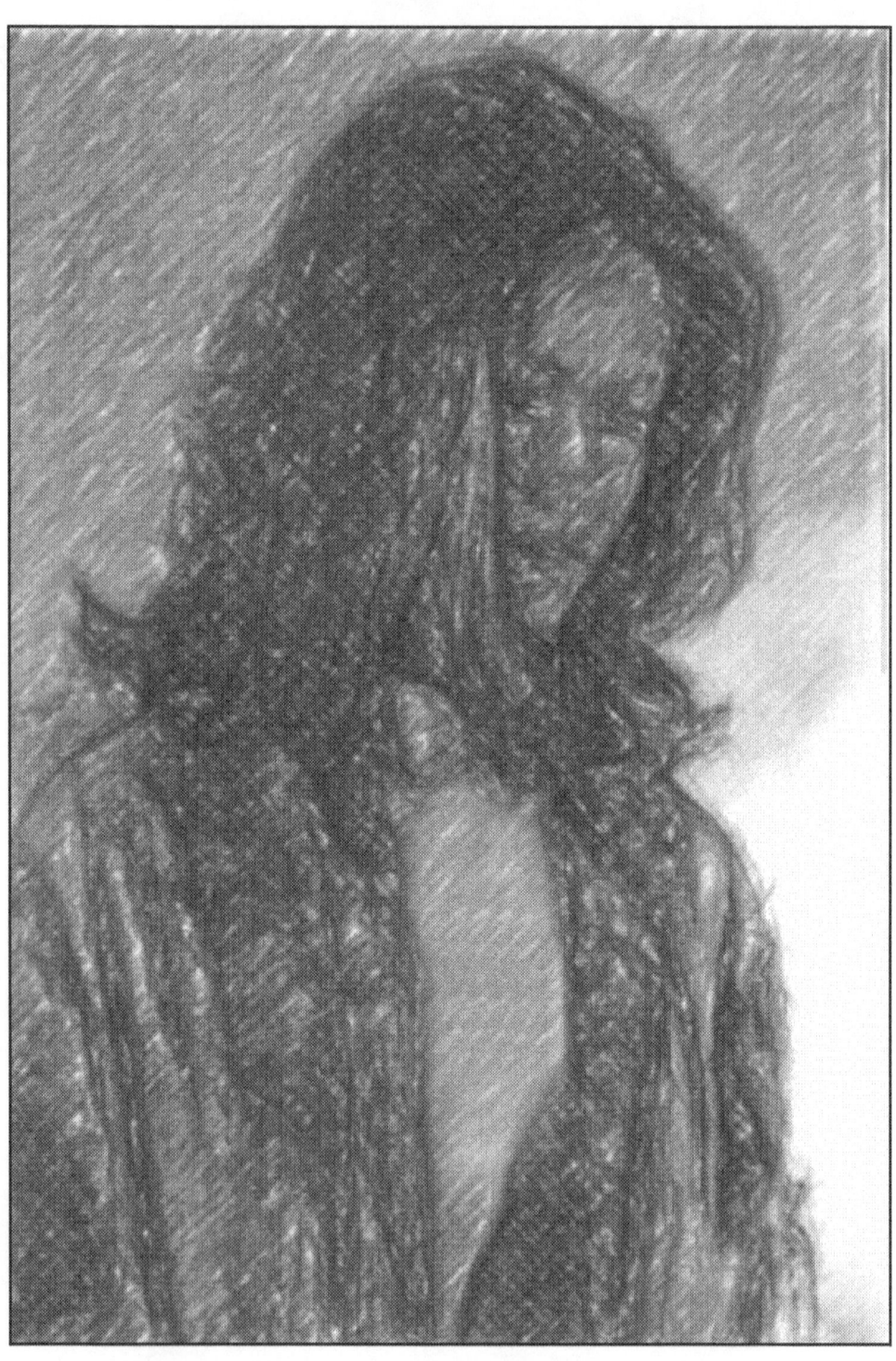

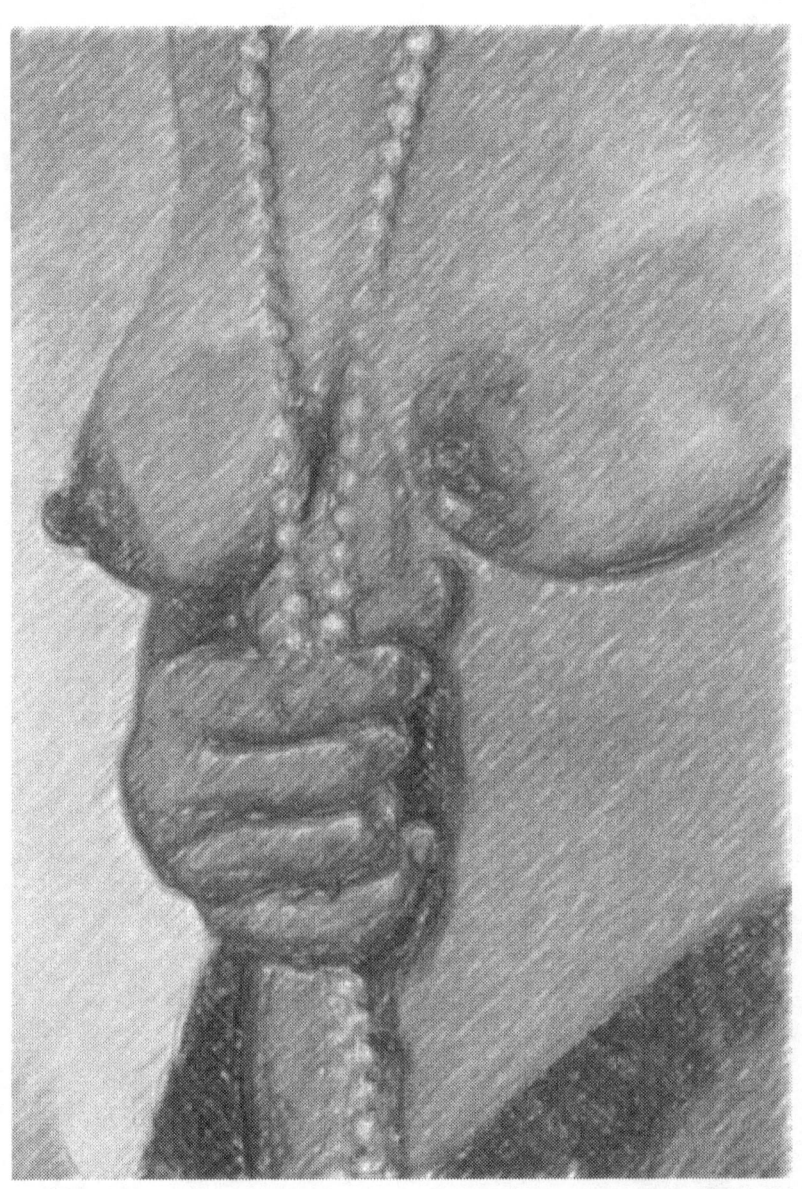

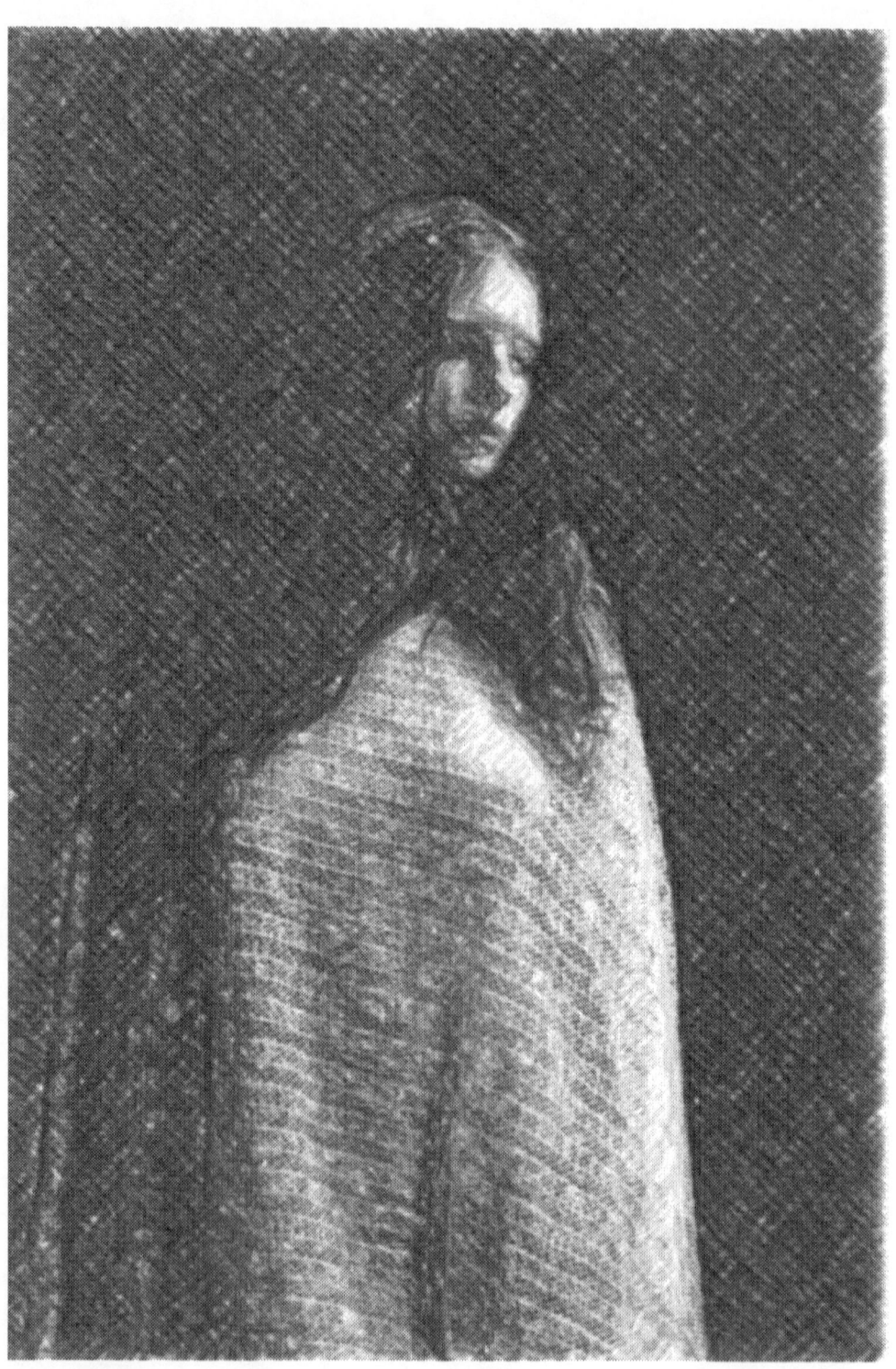

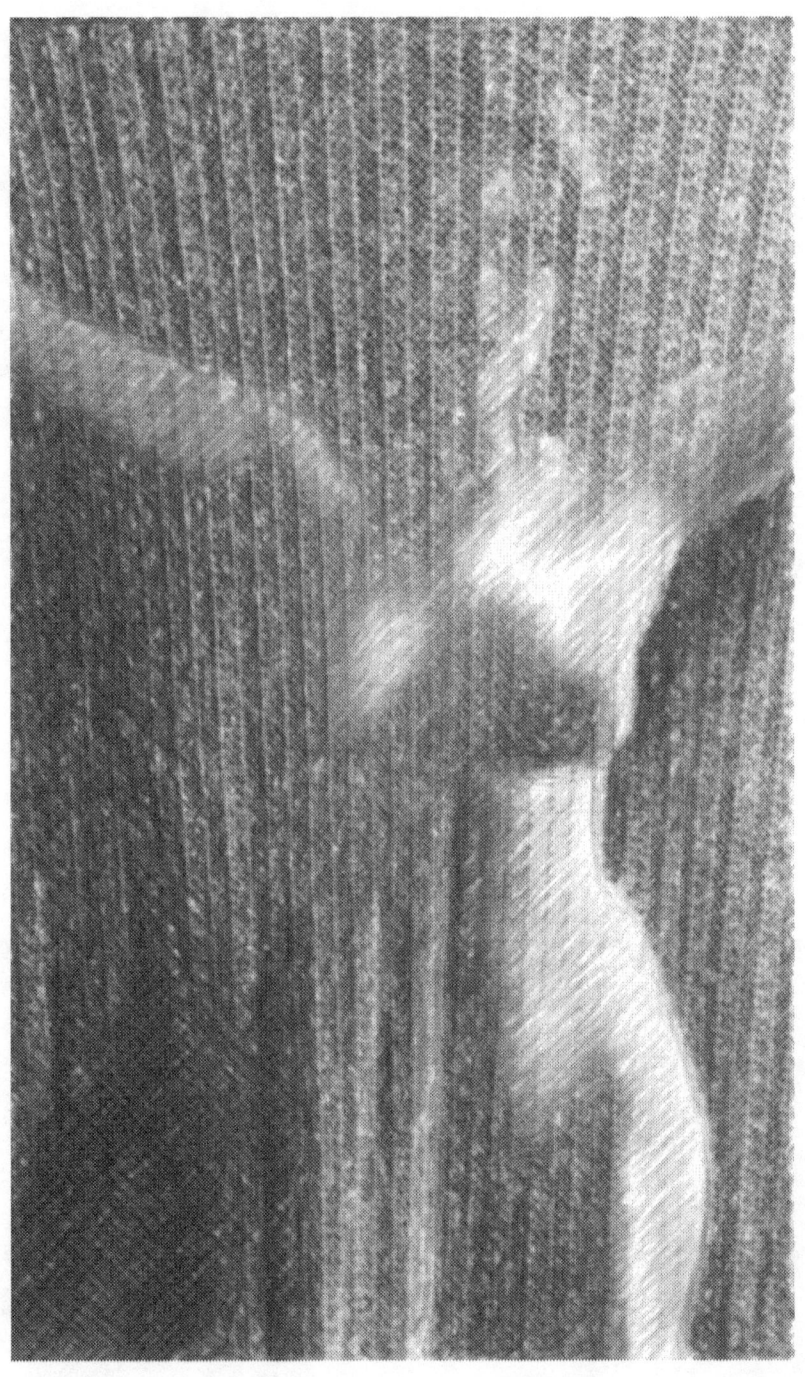

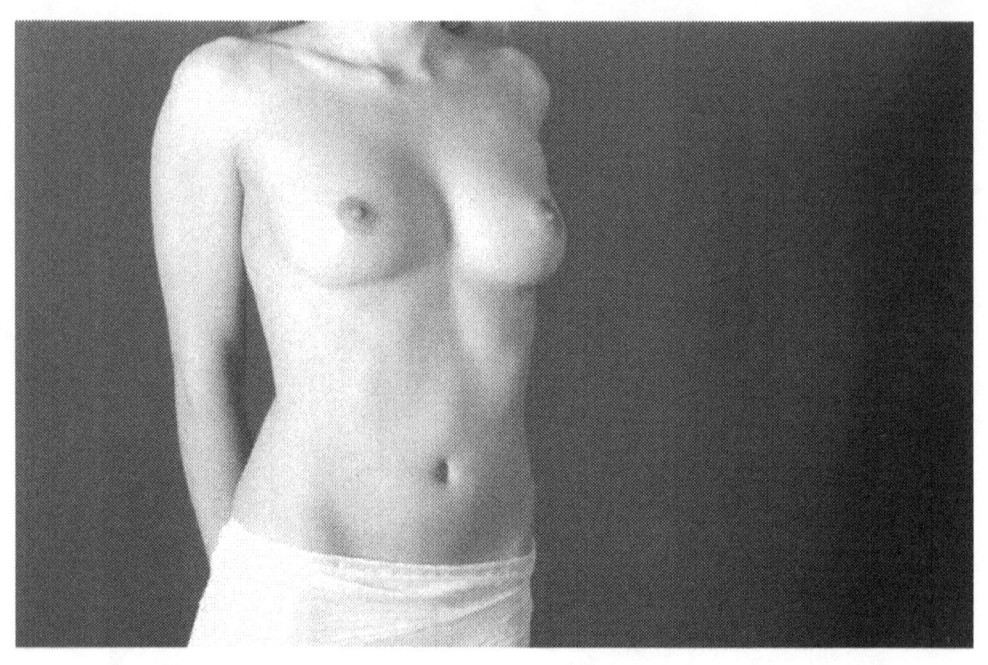

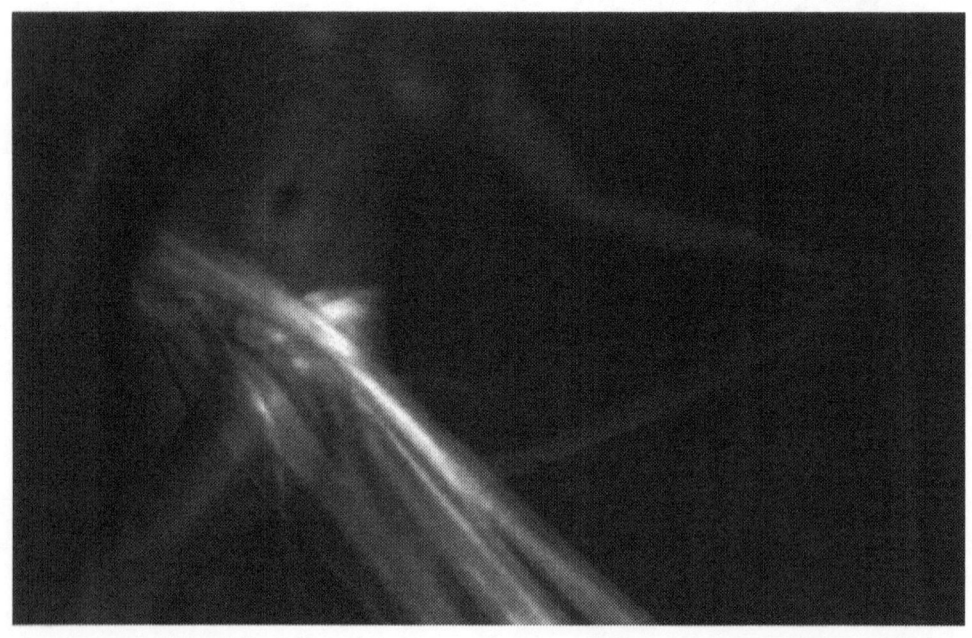

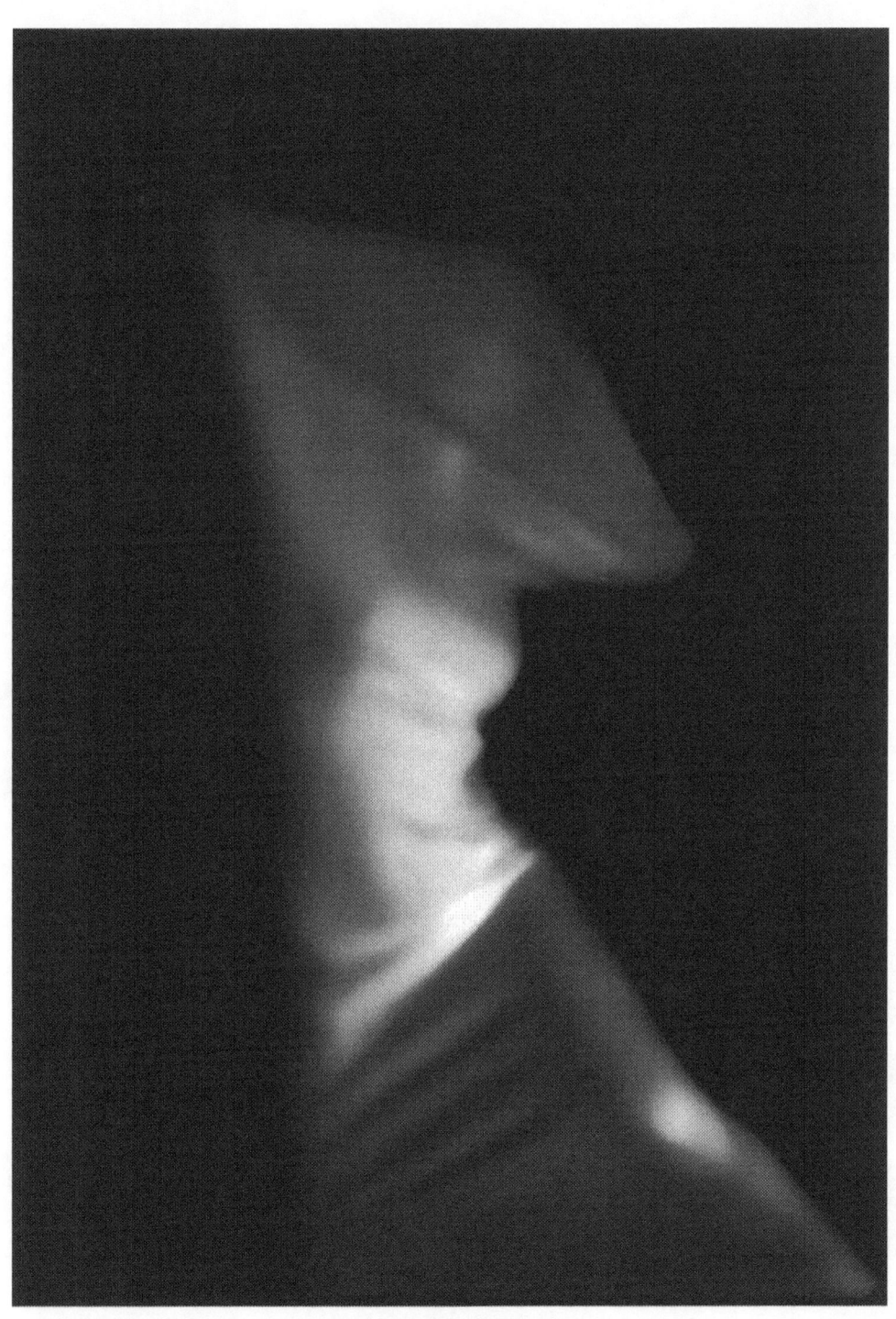

David Mecey

When I was asked me to impart some knowledge on choosing models as well as talking about posing, I began to run through my mind what it is that draws me to one model over another. For the most part, most women who become models are all fairly attractive, so physical beauty must be only a part of the process for me. The intangible is how well they relate to me, i.e., friendly without being too flirty, and then the other side of the coin, not so guarded in their mannerisms toward me that they appear uneasy.

I expect them to be excited about the shoot, not about me. I feel some photographers tend to get too caught up in the attraction thing, as that is what they feel makes for a "mood" between them during their shoot. In some cases, maybe, but for me that's a limiting factor as not every model you meet is going to be attracted to you in a guy/gal way. It's usually more so about your reputation as a photographer and much more about your work, itself. To me, your work is what makes or breaks a photographer when approaching and working with a new model whom you've never met.

Once that bond is secure between the two of you, then it's a combining of your talents, yourself as a shooter and she as a potential chameleon. Put them in the right combination to create a stunning photograph.

If those criteria are met, your job now becomes one of "seeing" not only the light but also her mood and body language in order to capture her beauty on film or pixels. I tend to watch how a model moves in real life to get a sense of who she is. If she's unaware of herself, totally natural and unguarded in her stance and sitting positions, I know she's going to probably be someone who will be very sexy, naturally, in front of the camera. Those who are more guarded usually need gentle guidance. I say gentle because you must move slowly to gain her trust so that the poses you're wanting are not putting her in what she may feel are compromising positions. This is very, very important.

Though, eventually, you will probably move to a place where her comfort level is much more giving. At that time you'll end up doing some wonderfully beautiful pictures without any drama or misunderstandings.

Remember, posing is something that should look stylishly normal. By that, watch how women stand or walk naturally when you are traveling or walking around public places. Watch how they sit, interact with one another, laugh, gesture, lean, etc. All of these moments should become a guidebook in your mind of poses. Use them as a starter kit when working with a model until she can begin to move naturally for you.

Opportunities for Growth

Once a shoot begins, it is at times difficult to stop and take a breath, to review what has be done. With film, it is not until a few days later that the artist can tell what is right or wrong with an image. Even when you are shooting with a digital camera and the results can be viewed immediately, you're not always looking at an image with a discerning eye.

In taking a look at these images you can, perhaps, see things you have done before: fingers gripping too tightly, a tense expression, the use of too much light, etc. Looking at them in this manner may serve to help you learn from my mistakes.

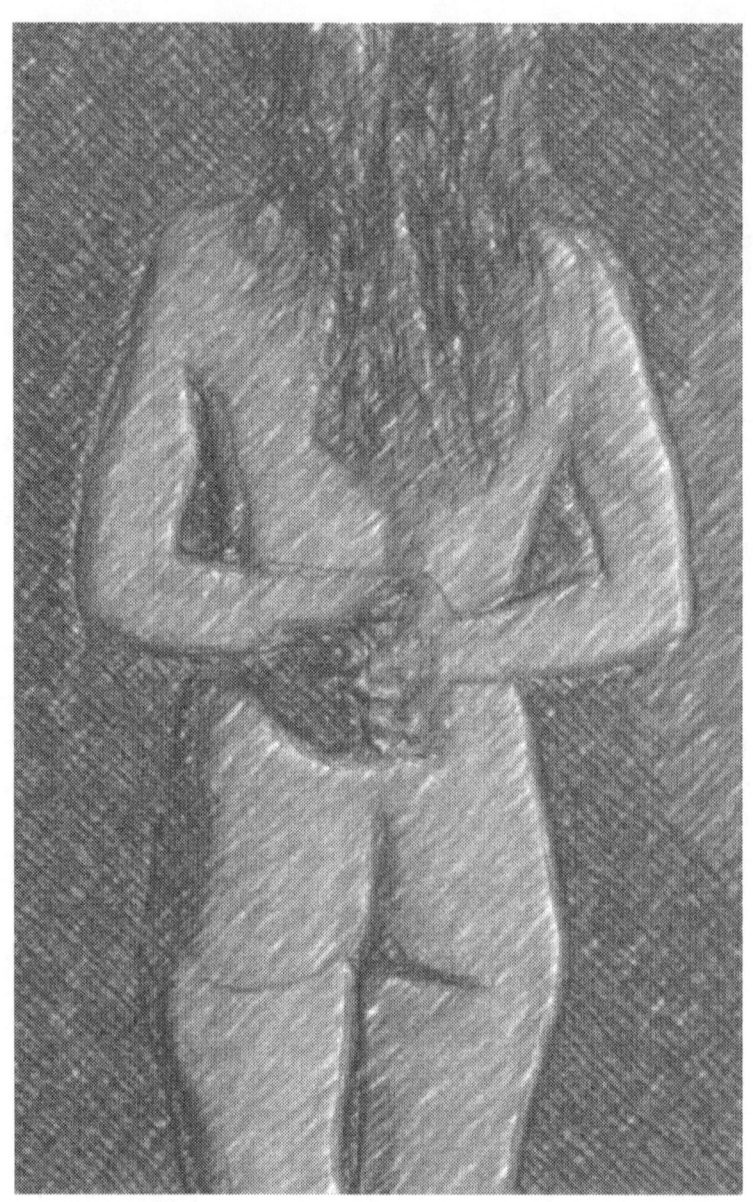

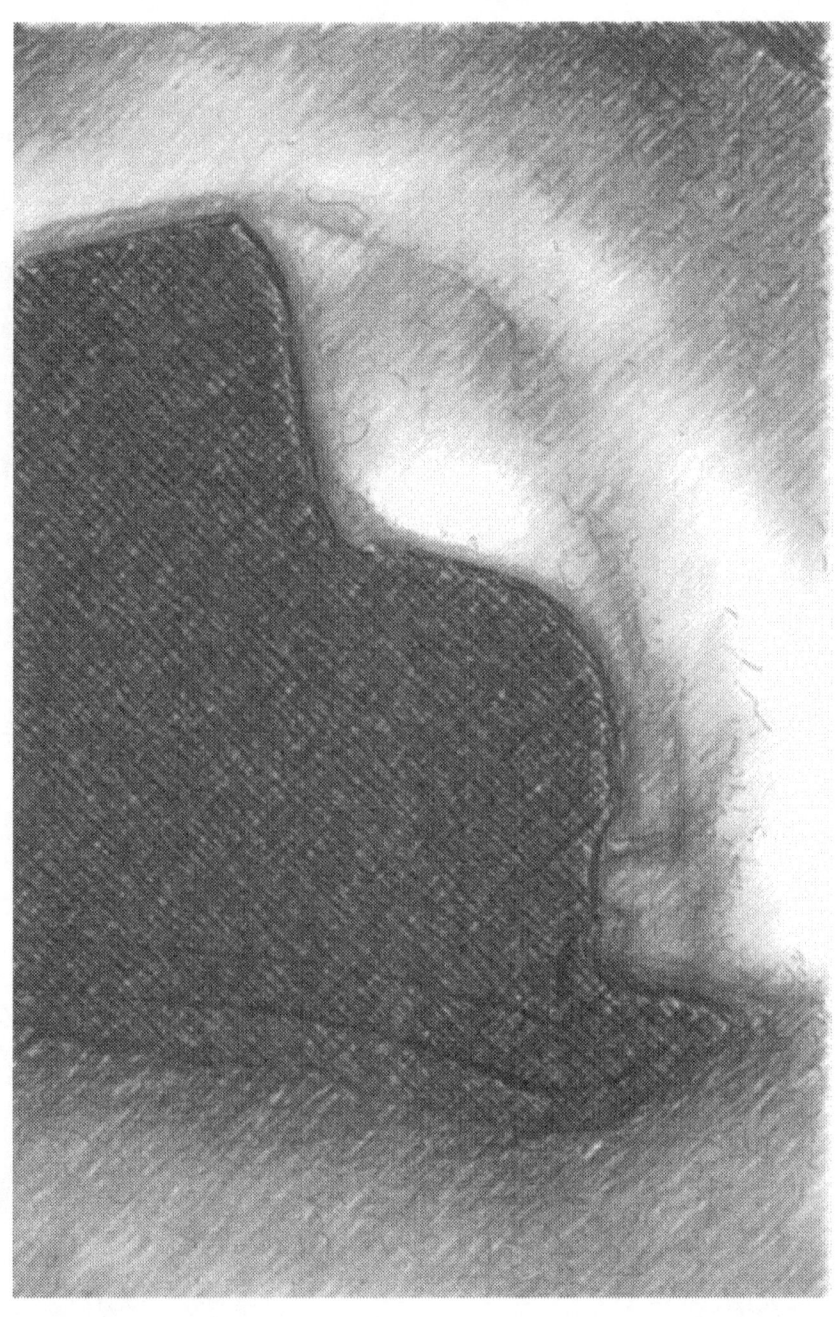

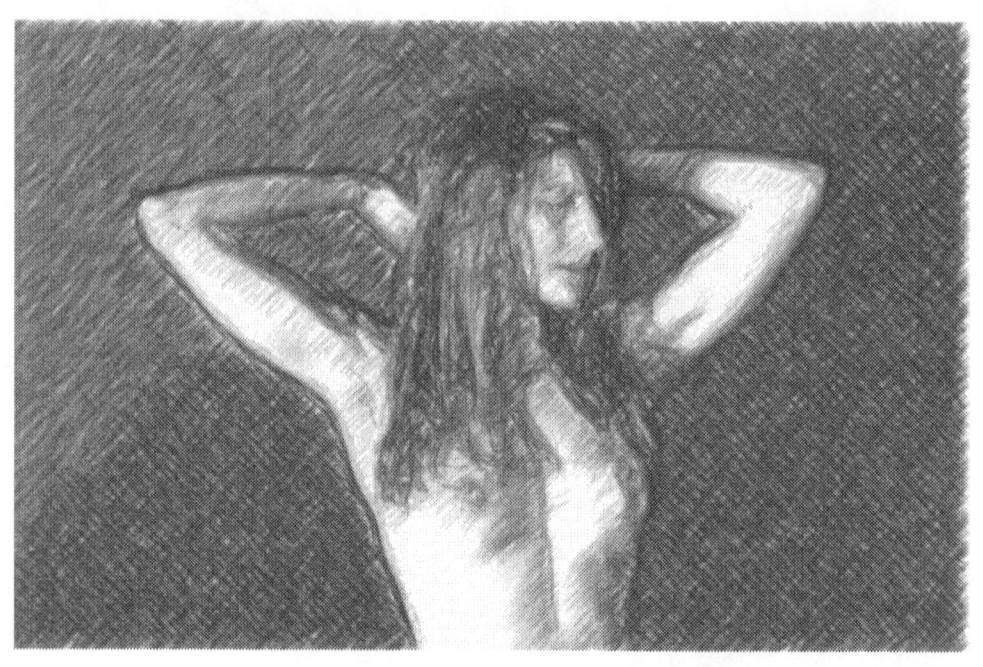

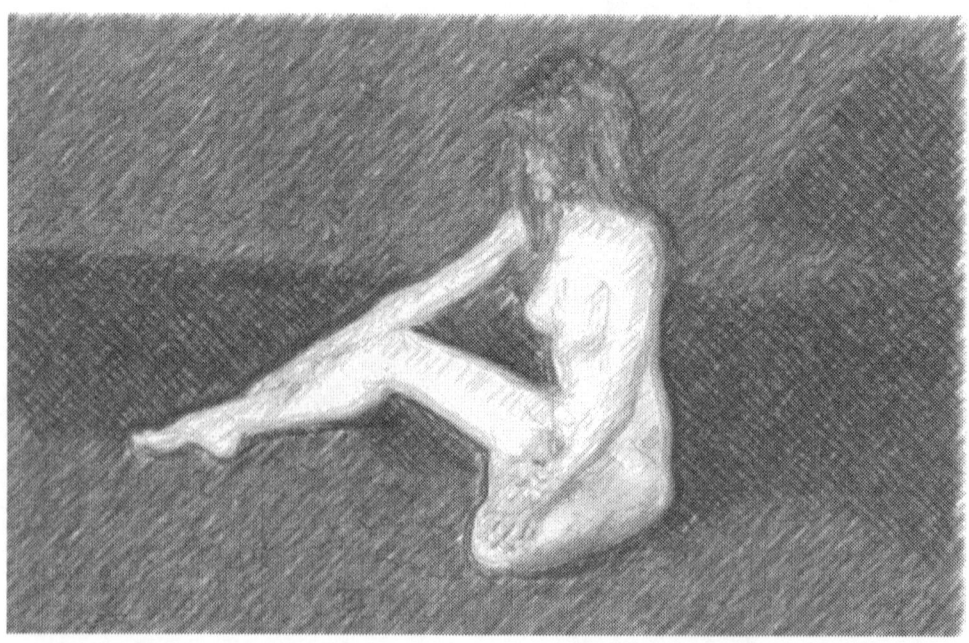

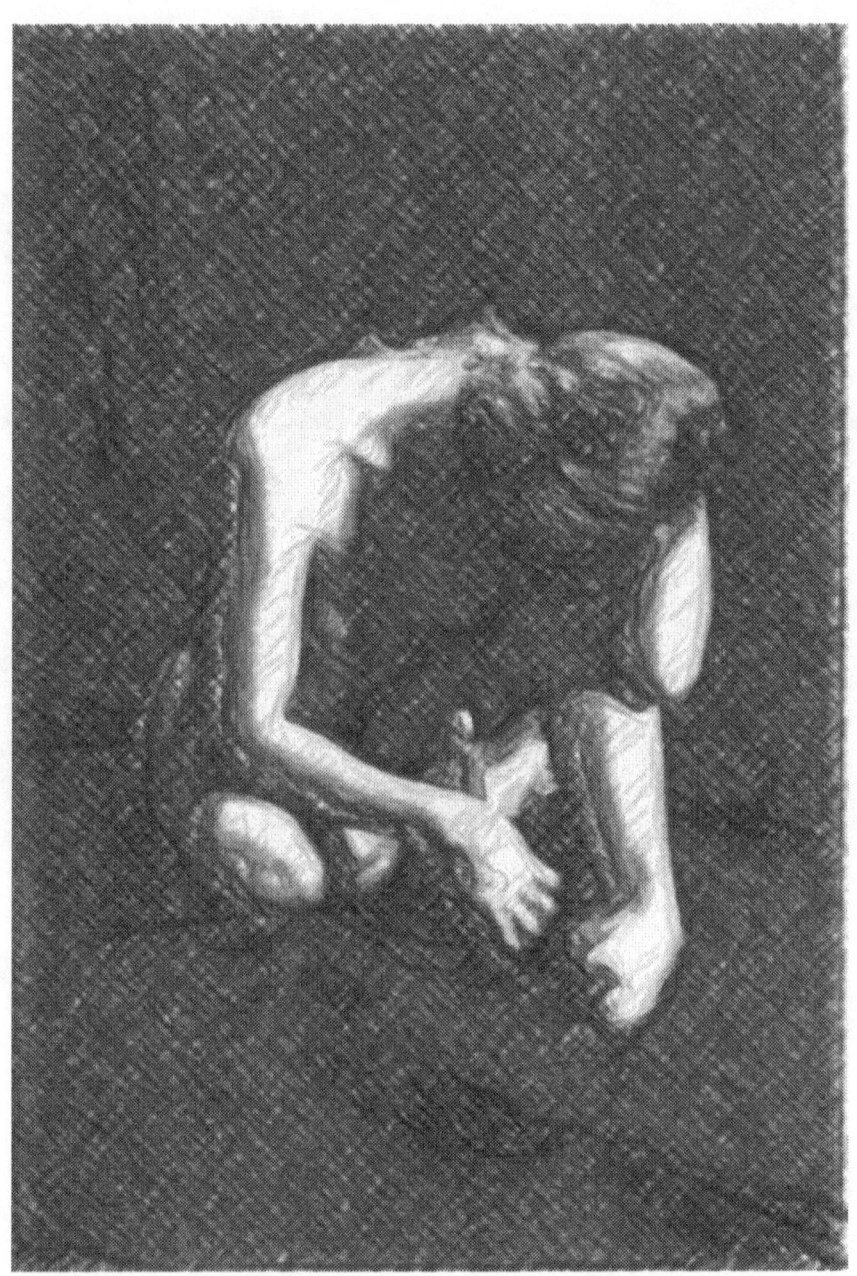

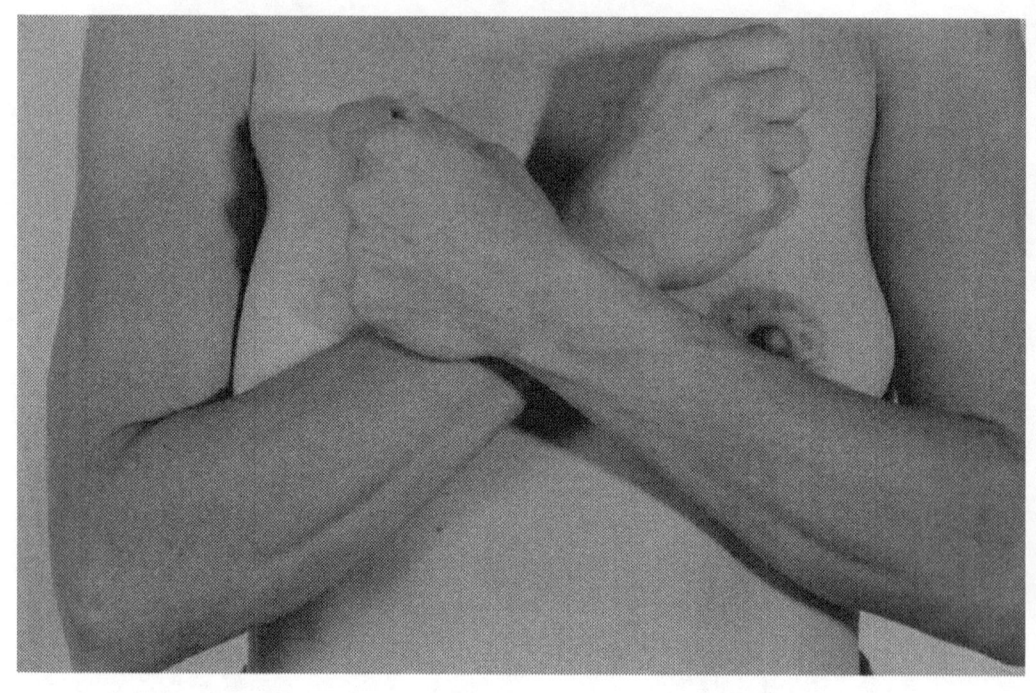

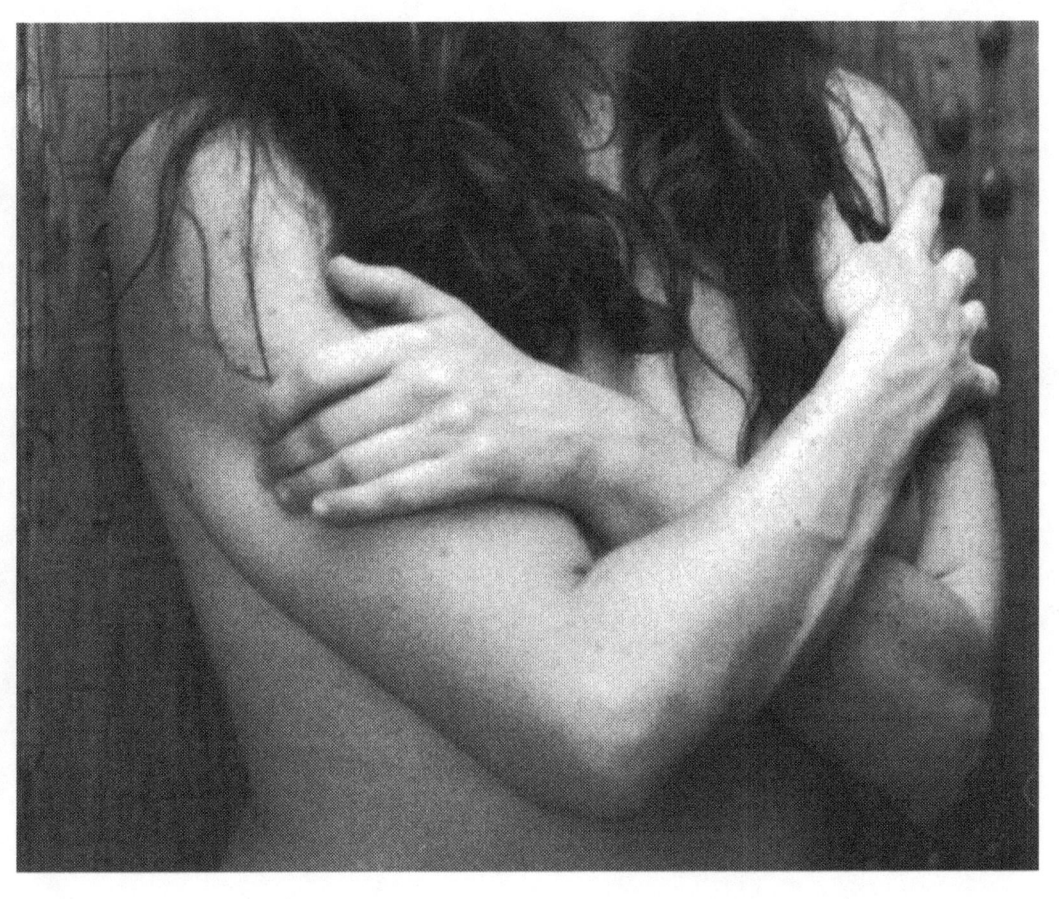

Contributors

From 1983 to 2004 **Steve Anchell** led the longest running workshop on photographing the nude in the world, *The Nude at Big Sur*. His book of the same title is a collection of his work made during that period in Big Sur, California. Prior to that, Steve worked with models, both clothed and nude, at his Hollywood, California studio. He continues to teach workshops in figure photography, both in the studio and outdoors, from his home in Salem, Oregon. Anchell is currently the editor of *Focus* magazine, and has written several books on darkroom processes, and is a contributing writer for *Rangefinder* and *Shutterbug*. (steve@steveanchell.com; www.steveanchell.com.)

Rose Bryan is a true professional and began fine art modeling while in college when she agreed to pose for her university's sculpture class. For more than ten years, she has worked with painters, sculptors and photographers with a focus on helping artists articulate, explore, and capture the images in their head. She is in much demand, and is the only model in a recently completed collaboration, "1X20," a photography project involving one model and 20 photographers. (rose.bryan@gmail.com, www.modelmayhem.com/pics.php?id=147947)

Known for her painted-gelatin silver photographs of botanical and figure studies, **Brigitte Carnochan** has been exhibited at galleries and museums nationally and internationally. *Bella Figura*: *Painted Photographs by Brigitte Carnochan*, containing 74 color plates of her botanical and figure studies, was published by Modernbook Editions in July 2006. A limited edition monograph with eleven original gelatin silver figures and flowers, *The Shining Path*, was published by 21st Publications, also in 2006. She has been featured in many publications and was a 2003 Hasselblad Master Photographer. (www.brigittecarnochan.com)

Jeff Dunas has been working commercially as an editorial, fashion, and advertising photographer since 1971. He has contributed photographic essays to publications worldwide, including *GQ, Life, Vibe, Entertainment Weekly*, and *Esquire*.

Living in Paris in the 80s, Dunas founded Melrose Publishing Company and Collector's Editions Ltd., a mail order distributor of fine-art photography publications. He has written and published in excess of 100 interviews with many leading photographers including Helmut Newton, David Bailey, Ralph Gibson, and Robert Mapplethorpe.

What Dunas is perhaps best known for, however, are his celebrity portraits, photo-essays and nude photography, for which he has garnered international acclaim and attention with his fine-art approach.

From a philosophy of "taking the pictures you want to take rather than that of the client's," came some of Dunas' most stunning series of photos and monographs, including his last three: *State of the Blues* (1998), a highly respected and awarded tribute to an American music tradition; *American Pictures* (2002), a poignant look at place, architecture and landscape in the place you were born; *Up Close & Personal* (2003), in which his portraits of celebrities serve to "try to find their unique edge, their magnetism." (www.dunas.com)

It took **Bill Lemon** 16 years from the time he was a part-time photographer until he decided to take the full-time plunge. Working in a studio at first, he decided that the confines of a studio were not appropriate to his photography style, and moved his studio to the outdoors in 2003. Lemon has published several books about nude and glamour photography, with a new one due out in 2008. Bill Lemon has shot more than a dozen covers for a variety of magazines, and has been the subject of many articles, in magazines such as *In Style*, *The Robb Report*, *Redbook*, *Shape*, *Shutterbug*, and *American Photo*. (bill@billlemon.com; www.billlemon.com)

Robert Farber's style has helped to influence a generation of photographers through vast public exposure of his work. His eight coffee table books have sold more than a million copies. Farber's fine-art photographs have been published in virtually every form, and his work has been exhibited in Japan and Europe, as well as the United States.

Robert Farber received the Photographer of the Year award in 1987 from the Photographic Manufacturer's Association. In 1995 he received the ASP International Award, given by the Professional Photographers of America, and The American Society of Photographers. This award has been given to those who have made a significant contribution to the science and art of photography. Some previous recipients of this award include Dr. Edwin Land (inventor of the Polaroid), George Hurrell and *National Geographic*.

Farber's involvement in the Internet began in 1994 with Farber.com, a virtual gallery of his work. Because of its popularity, Farber created Photoworkshop.com, which has become the most successful and unusual way to learn photography for both professionals and amateurs on the internet. (www.farber.com; www.photoworkshops.com)

Since the late 70s, most of the photographic career of **David Mecey** has been focused around the work he has done for *Playboy*, where he became known as one of the finest photographers for shooting beauty in the world. But he believes there comes a time in a photographer's life when "you seem to yearn for those times when you could just shoot for the sheer fun of it. Not a publisher to answer to, nor an art director, no editor, no associate editor, not a single soul save one. You. And that's what he has tried to create with his black-and-white project, tentatively titled, *Passion*, which features models he has known and met during his career.

Mecey served for over two decades as staff photographer and contributing photographer to *Playboy*, where he found 20 Playmates of the Month, shot 6 gatefolds, and worked on multiple pictorials. He has shot over 40 magazine covers for various publications as well as photographed a number of celebrities. (www.davidmecey.com)

Mark Nelson is the author of *Precision Digital Negatives for Silver & Other Alternative Photographic Processes*, often recognized as the authoritative book and system for creating hand made prints from contact printed digital negatives. He has helped some of America's leading contemporary photographers develop turnkey systems for contact printing hand made, museum quality editions with digital negatives—utilizing a variety of alternative processes.

Nelson has also contributed articles for several magazines and a chapter on digital negatives for *The Platinum & Palladium Print* by Dick Arentz. He is currently consulting regarding the use of the Precision Digital Negatives Process to enhance the polymer plate photogravure process. His personal work, which can be viewed at his website, includes figure, still life, and landscape work, hand printed with the platinum/palladium process. Currently, he represents himself for sales of his personal artwork. (ender100@aol.com , www.MarkINelsonPhoto.com, www.PrecisionDigitalNegatives.com)

Elizabeth Opalenik privately conducts photography workshops internationally and in partnership with Santa Fe Workshops, Maine Photographic Workshops, the British Guild of Portrait Photographers, National Geographic Expeditions, The Recontre d'Arles, and others.

For the past 20 years she has returned to Provence and Tuscany conducting workshops on the figure, travel and alternative processes. Her current workshop, "Imagination & Dreams," is one of the most insightful and inspirational photography workshops, and are continually sold out. Her first monograph, *Poetic Grace*, is set for release in the summer of 2007. (www.opalenik.com)

Ashley Rae began modeling for artists in 2002 and felt like she had discovered a new "me." Enjoying exploring her artistic side, and collaborating with photographers and artists to create artwork that evokes emotion, the Colorado resident soon decided to become a professional model.

She is much sought-after now and maintains a presence on One Model Place (www.onemodelplace.com/AshleyRae), a very popular meeting place for models and photographers.

Kate Taylor has been modeling for almost two years and has already amassed a formidable portfolio. She brings a wealth of experience in the performance arts to her work, and travels extensively, enjoying working with a wide variety of artists and photographers all over the world. (www.katyt.com)

Kim Weston has been a fine-art photographer for 30 years, and is a third-generation member of one of the most important and creative families in photography. He learned his craft assisting his father Cole in the darkroom making gallery prints from his grandfather Edward's original negatives. Kim also worked for many years as an assistant to his uncle Brett, whose bold, abstract photographs rank as some of the finest examples of modern photographic art.

For six years Gina and Kim Weston have been selling photographs by Kim, Edward, Brett and Cole Weston on their website www.kimweston.com. Weston Photography also has a private gallery located at Wildcat Hill, Carmel Highlands, California.

References

Aperture. *Edward Weston Nudes.* 1977.

Ashford, Rod. *Lighting for Nude Photography.* RotoVision, 2002.

Bernhard, Ruth. *The Eternal Body.* Chronicle Books, 1986.

Biancani, Laurent. *Nude Photography, The French Way.* amphoto, 1980.

Braham, Phil. *Naked Women.* Thunder's Mouth Press, 2001.

Clergue, Lucien. *Practical Nude Photography.* Focal Press, 1984.

Cornell, Peter-Richter. *Nude Photography.* Prestel, 1998.

Farber, Robert. *Farber Nudes.* amphoto, 1983.

Garduno, Flor. *Inner Light.* Bulfinch, 2002.

Hedgecoe, John. *Nude and Portrait Photography.* Simon and Schuster, 1985.

Hope, Terry. *Nudes, Developing Style in Creative Photography.* RotoVision, 2001.

Kelly, Jain (editor). *NUDE: THEORY.* Lustrum Press, 1979.

Lorenz, Richard. *Imogen Cunningham, On the Body.* Bulfinch, 1998.

Olley, Michelle. *Venus: Masterpieces of Modern Erotic Photography.* Thunder's Mouth Press, 2000.

Stebbins, Thomas E. Jr. *Weston's Westons, Portraits and Nudes.* Bulfinch, 1989.

Trampe, Stan. *Black & White Nude Photography.* Amherst Media, Inc. 1998.

About the Author

A professional photographer and writer for more than 25-years, Tim Anderson has photographed and interviewed many performers, politicians, sports figures, and photographers, but will always say that his favorite area of photography is fine art portraits, particularly the nude. He lives in the Southwest and is currently the publisher and managing editor of *CameraArts* magazine.

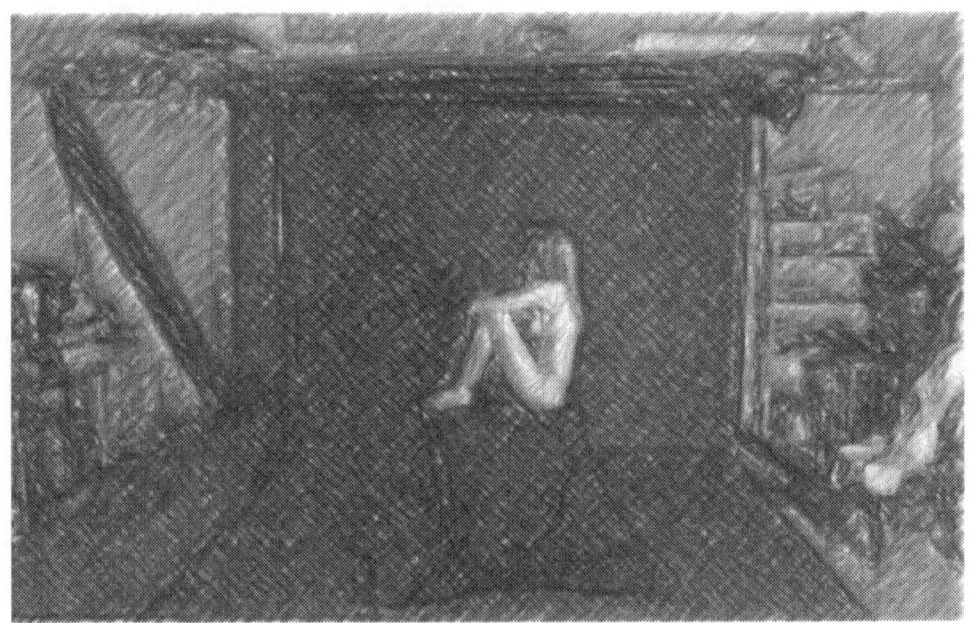

Acknowledgements

As you know, no book can be completed without the help of many people. In the developing stages of *Poser*, support came in many ways: encouragement from friends, kudos from peers, and an idea that wouldn't let go. Special thanks go to the contributors: Lucien Clergue, for writing the Foreword, Steve Anchell, Rose Bryan, Brigitte Carnochan, Jeff Dunas, Robert Farber, Bill Lemon, David Mecey, Mark Nelson, Elizabeth Opalenik, Ashley Rae, Kate Taylor, and Kim Weston.

Order Form
Poser

A sketchbook of ideas for artists and models

(Please print or write legibly)

Name: _____

Address: _____

City: _____

State: _____ Zip: _____

Phone: _____

Email: _____

Copies: _____ Gift: Y: ____ N: ____

(Gift for)
Name: _____

Credit Card Info:
MC ____ Visa: ____ Date of Expiration: _____/_____

Card Number: _____

CVV (three digit number on back of card) _____

Price: $24.95/includes shipping
Please include a tax of 6.75% ($1.50) if ordering from New Mexico.

Send check or card information to:
Cygnet Press
P.O. Box 3941
Albuquerque, NM 87190
(info@cygnetpress.com)

clip
or
copy

Notes:

Notes:

Notes:

Notes:

Notes:

Notes:

Notes:

for Annie..............